Beyond the T-Test

Scott A. Pardo

Beyond the T-Test

Springer

Scott A. Pardo
Global Medical & Clinical Affairs
Ascensia Diabetes Care
Valhalla, NY, USA

ISBN 978-3-031-84478-2 ISBN 978-3-031-84479-9 (eBook)
https://doi.org/10.1007/978-3-031-84479-9

This Springer imprint is published by the registered company Springer Nature Switzerland AG
The registered company address is: Gewerbestrasse 11, 6330 Cham, Switzerland

Preface

This book was written with a certain group of people in mind. Those people are professionals who are not statisticians but who must review, discuss, and evaluate statistical methods used for various applications. In particular, people involved in regulated industries, such as pharmaceuticals and medical devices, are inundated with applications of statistics, and they are often required to either justify or explain those methods. Regulatory reviewers must assess whether the methods applied were appropriate and applied correctly.

There are of course others who also must judge, explain, or even recommend the use of statistical methods. They too could benefit from this work.

Finally, students of biological, social, or physical sciences could benefit from this as a text on a second course in statistics.

Many people who study many disciplines have taken at least one college course on elementary statistics. But just like the high-school Euclidean geometry you were taught, without continual use, the knowledge falls away from you like waves from the beach. Then, many find themselves in some academic or professional capacity faced with the dilemma of needing to understand, interpret, or even justify some statistical analyses. Sometimes these unfortunate individuals attempt to consult their old elementary statistics text, only to find that, well, it just doesn't cover the material they are presently struggling to comprehend.

This book is an attempt to bridge that gap, between the old elementary statistics course and the need to assimilate and process information about statistical analyses that seem to go beyond their previously acquired (and lost) statistical knowledge. It turns out that many such individuals may have even had an education in mathematics commonly presented to undergraduates in sciences and engineering. But, that knowledge may have also slipped through the cracks of their gray matter, and in fact, it probably wouldn't have helped with their current issues.

So this book will hopefully awaken those sleepy memories that have settled deep in the dark corners of your mind, and maybe kindle an interest, now that you have been faced with the task of trying to understand those arcane notions. The fact is, there is nothing mystical about statistics. It just takes thinking about the universe in a way you may never have been taught.

It is my opinion that a modicum of mathematics should be included in a discussion of statistical methods. I realize that "modicum" is a very subjective term. Therefore, I have included an appendix that contains all the mathematics you would need to have a sufficiently deep understanding of the statistical material in this book. In fact, the mathematical concepts presented in the appendix only presume a rudimentary familiarity with high-school algebra. The appendix has material that goes through elementary calculus and matrix algebra, but fear not! You will not need to solve difficult mathematical puzzles, prove theorems, or put "word problems" into some mathematical context. In fact, without any attempt to garner the mathematical skills, you can still get much out of the material in this book.

The examples involve the use of computer programs in the form of R scripts. For those not familiar with R, it is a computer programming language specifically designed to compute statistical things. All the code used in this book is included as a second appendix and is organized by chapter. There are a number of books on the subject of R programming. The more adventurous readers may endeavor to learn enough about R to actually run the code in the appendix. The data sets and code are also available online via the publisher, provided you actually purchased the book new. In the appendix, each script (program) is organized by chapter and has an underline to mark its end. The underline is not part of the code. The programs all begin with the statement, or R function, "setwd." Each script has an identifier of the form "Code xx.y," where "xx" is the first chapter number in which the code was utilized. In the text, the output will have a pointer to the code used to generate it.

This book could be used as a text for a second course in statistical methods for non-statisticians, from psychologists and biologists to sociologists and physicists. In fact, this book could be useful for anyone whose academic or professional endeavors would lead them to working on assimilating and interpreting statistical documents, so that at the very least they may be able to formulate better questions to the writers of those documents.

OK, so have at it. You've got this. I hope it will enlighten you and give you a sense of accomplishment and independence.

One other thing. This book does not contain the entire universe of statistical concepts, methods, or analyses. It is a survey of many methods I have personally encountered in my academic and professional experience. I hope that you find something here that strikes a chord of familiarity. If there is something you have not seen in this book but you have encountered yourself, I encourage you to research it and try to understand it.

Valhalla, NY, USA Scott A. Pardo

This work contains media enhancements, which are displayed with a "play" icon. Material in the print book can be viewed on a mobile device by downloading the Springer Nature "More Media" app available in the major app stores. The media enhancements in the online version of the work can be accessed directly by authorized users.

Acknowledgements

There are many people who deserve thanks and gratitude. First, I would like to thank the regulatory affairs team from Ascensia Diabetes Care:

Brian Truchan
Shahrir Alam
Robin Martin
Sangram Yadav
Divya Munjpara

The questions these dedicated and intrepid workers have posed to me over the years inspired me to write this book.

I would also like to thank Rosanna Cirio, my supervisor and colleague. She has also provided me with many questions and insights I would never have found on my own.

I must thank my colleague, Bern Harrison, whose keen insights and unrelenting adherence to facts have kept me honest. To say you are a Jack-of-all-trades does not do you service. One thing I am especially impressed with is your ability to make computer simulations. Professor Moriarty owes you, Sherlock.

I also want to acknowledge and thank Dr. Michael Pardo, Dr. Yehudah Pardo, Dr. Jeremy Pardo, Dr. Anna Pardo, and Dr. Rachael DeTar. Your questions have taught me so much. I love you all.

Finally, I can only try to thank the one who has been my guiding light, from writing all those computer programs that brought my Ph.D. dissertation to life, for encouraging me not to give up on finishing my Ph.D., through all the trials of my working life, through writing this and other books, and through all the joys and tears that life brings. You have contributed many ideas and questions that have enhanced my writing. Of course, I can never repay you for all you have given to me. Te amo, Palomika mia.

Competing Interests The author has no competing interests to declare that are relevant to the content of this manuscript.

Contents

Chapter 1
Populations, Samples, Parameters, and Statistics

A population is the total number of individuals about which the experimenter is interested in knowing something. Individuals will be referred to as experimental units. Experimental units are the entities upon which a measurement or observation is made. Experimental units can be single organisms, parts in a production line, or entities such as classrooms full of students, platoons in the Army, herds of animals, or agricultural fields. So, a population might be all the elephants living in Botswana, or all the fourth grade students living in Davenport, Iowa at some particular time. It could be an infinite population, such as all stars in the entire universe, or finite, such as all the bowling pins in a particular bowling alley. It could even be a small population, say all the jellybeans in a particular bag. The point is that the population is what is of interest. Often, only one particular aspect of the population is of interest, such as the eye color of all those fourth graders, or the weights of all the elephants. The aspect (or aspects) in which the experimenter is (are) interested are generally described by some variable, which we will call a random variable. These random variables are always represented by numbers. Those numbers can be values on a continuum, such as weight, or integers, such as the number of fourth graders with blue eyes. Random variables whose values are decimal numbers are called continuous. Random variables whose values are integers are called discrete. Sometimes the items in the population are actually collections, such as cars on a particular train going past a particular crossing over some particular time interval, and the random variable might be the number of cars per train. The important characteristics is that the population is the total of all items that are of interest.

Populations have parameters that describe the frequency distribution of random variables measured or observed on individuals in the population. For example, the population of elephants in Botswana has an average weight of all the elephants. The number of cars per train traveling past a particular crossing within some particular time interval has some average. There are many other possible parameters, such the 95th percentile or 5th percentile. In the case of a bag of jellybeans, we could talk about the random variable that is the number of red jelly beans, or the percentage of

S. A. Pardo, *Beyond the T-Test*, https://doi.org/10.1007/978-3-031-84479-9_1

red jelly beans. These numerical values are called parameters; they describe something about the frequency distribution of random variables.

In the case where the population has relatively few individuals, such as a single bag of jellybeans, it may be possible to easily obtain the value of parameters. But what if the population is really large or it is just not feasible to obtain all the individuals who belong to that population? In the case of Botswanan elephants, while there may be a finite number of individuals (at least within some fixed time interval), it may not be practical to weigh every single individual. So, we can only guess at the values of parameters, such as the average weight of individuals. We make guesses like this by sampling the population, and measuring or observing the individuals in the sample only. So, perhaps we obtain 27 elephants, and weight them. Can we calculate their average weight? Certainly. This average, which is based only on the sample individuals, is called a statistic. The statistic is an estimate of the population parameter. Suppose the jellybean bag had 100,000 jelly beans in it. Maybe we could not observe every one of those jellybeans. So, maybe we take a sample of 107 jelly means, and count up the number of those 107 which are red. We could then compute the percentage of the sample jellybeans that are red. This sample percentage is a statistic and is an estimate of the population parameter.

So, to recap, we have populations that consist of all the individuals of interest. The population could be finite in size, or even infinite (like all the elephants ever born in Botswana, or all the children who were, are, and will be 4th grade students in Davenport). Random variables are numerical values that are measured or observed on individuals. These random variables are associated with parameters, numerical values that are in some way related to the frequency distribution of the variables. Often it is not feasible to measure or observe every individual in a population, so we sample, and make measurements or observations on those individuals in the sample.

There are many ways to obtain a sample. A random sample is one in which every individual in the population is equally likely to be selected. This is the simplest form of sample, that they type which will mostly be assumed for much of this book. However, random samples may not be the most representative. For example, elephants come in different sexes and ages. Sometimes proportional sampling is done, when there are known to be subsets of the population that differ from each other, but the experimenter still wants to estimate some overall parameter. Other forms of sampling will be discussed, but not now. For now we will assume there are no a priori known subsets of the population that somehow differ from each other.

The arithmetic average is a fairly common and widely familiar statistic. The formula is fairly simple. Suppose the sample consists of n individuals, and we measure random variable X on each one. So, we will call the sample values x_1, x_2, … x_n representing the measurements of variable X on each of the n individuals (the order of those individuals is not important in this case). So the arithmetic average can be described mathematically as:

$$\overline{x} = \frac{x_1 + x_2 + \cdots + x_n}{n} = \frac{1}{n}\sum_{i=1}^{n} x_i$$

The use of the capital Σ is used in statistical formulas as a short-hand for summing a set of numbers. Likewise the capital Π is used to mean multiplying (product) a set of numbers.

The formula:

$$\frac{1}{n}\sum_{i=1}^{n} x_i$$

is an estimate of the population mean. Clearly, if we got two different samples of size n and calculated the arithmetic means of each, we would most likely not get the same numbers. Yet, each of those two arithmetic means are estimates of the population mean. Thus, the sample statistics is not the population parameter; it is only an estimate, and as such has some error associated with it.

Mathematical statistics is concerned with, among other things, finding formulas, called estimators, for estimates that minimize the error.

At this point, we are simply making the points that:

1. Populations are the total set of individuals who may be of interest; that set could be finite or it could be infinite;
2. Samples are subsets of individuals from a Population; they can be selected in a number of ways, but the one which will mostly be assumed in this book is the random sample;
3. Populations have parameters that relate in some way to the frequency distribution of some random variable;
4. Random variables are things that are measured or observed on individuals in a population; their values are always represented by numbers;
5. Statistics are estimates of Population parameters, they are based on the resulting values of random variables measured or observed on the individuals which comprise the sample;
6. There are two classes of random variables; discrete and continuous; discrete variables are those whose values are represented by integers; continuous variables are represented by decimal numbers.

There is an underlying and overarching assumption throughout this book. Namely, that we sample because it is either impossible or impractical to see ever individual in the population of interest, and we do not actually know the values of any population parameters. Thus, can only get sample statistics, and never the actual value of parameters. We may guesses about the interval in which the parameter value lies, or be may be able to say that the parameter is not likely to equal some particular value. This is to many a foreign, nay, alien concept. From a very early age we are taught that gravity is approximately 9.8 m/s squared, that force is equal to mass times acceleration, that an inch is 2.54 cm. These are all well-established facts (even though the acceleration of gravity had to be determined empirically, at least initially). But it you want to know how likely a vaccine will protect against some particular disease, or the average torque required to unscrew the lid off a container, or the average current draw on a pacemaker battery over its life, at best you can estimate it with sample data by calculating a sample statistic.

The scheme by which data are to be gathered, including the number of samples, calculation of sample statistics, the estimation of parameters, and the assessment of how trustworthy those statistics might be for estimating population parameters, are fundamental goals of the science of statistics.

Keep in mind that sample data provide some information about the population, but that information is wrapped in a cloak of probability. Two samples from the same population may give you different information. Since experimenters can generally not observe the entire population of interest, the information gleaned from the sample data must be couched in the language of probability.

1.1 Summary

A population is the sum total of all possible experimental units of interest. A sample is a (usually relatively small) subset of units from the population. The point of statistics is to use the data from the sample to estimate parameters of the population and make inferences about those parameters. In many cases, there are formulas, called estimators, which use sample data to compute the estimates.

1.2 Points to Ponder

1. A sample of 5000 lemmings were weighed. Their average weight was 1.6 kg. Is this a population parameter or a sample statistic?
2. Name three random variables that involve household items.
3. Are there conditions under which the mean of 200 observations of random variable X would constitute a population parameter.

Chapter 2
Some Probability Concepts

2.1 Random Variables

The primary probability concept we will use is called a random variable. A random variable is actually a function. It is generally sufficient to think of a random variable as a number whose value cannot be predicted exactly before any observation is made. So, gravity is not usually thought of as a random variable. On earth, gravity is 9.8 m/s^2, roughly. We do not need to measure it now in order to know it. However, the flight velocity of the laden European swallow is not a known value, per se. Some swallows may be faster than others, even when laden with the same weight and volume.

Random variables have ranges of values, and the ranges are associated with probabilities. So, for example, tensile strength of some object is a force that may have some variation from part to part, even though the parts are essentially "identical." Thus, we cannot say a priori exactly what the tensile strength of a particular part might be. Rather, we might be able to say that the tensile strength of a particular part will fall between 1.000 and 1.010 lbs. with a probability of 82%. Thus, the part may have a tensile strength of 1.003 lbs., but we may only know that its tensile strength will fall between 1.000 and 1.010 with probability 82%.

Random variables come in two basic flavors; continuous and discrete. Continuous random variables take on values that can be described by decimal numbers, such as tensile strength, length, electrical current, etc. Discrete random variables taken on integer values.

All random variables, continuous or discrete, are associated with a function that describes any probability one might like to know. Unfortunately, the exact form of this function, called the cumulative distribution function, or CDF (Meyer, 1970), is almost always unknown. One of the jobs of statistics is to make a guess about the form of the CDF, using observations, or data. The process of making guesses using data is called estimation and inference. More on that later.

S. A. Pardo, *Beyond the T-Test*, https://doi.org/10.1007/978-3-031-84479-9_2

Often we put some structure on the form of the CDF for some random variables. Structure means a model. The model is a mathematical representation of the CDF, with constants, or parameters, whose values can be estimated using data. Such functions will be referred to as "parametric distributions." The parametric distributions form families, whose mathematical forms are the same, but whose parameter values vary from one particular situation to another. Parametric distributions are often described by mathematical functions that are actually the first derivative of the CDF. These functions are called probability density functions. For discrete random variables, the density function is sometimes referred to as the probability mass function.

In the case of continuous variables, the probability that a variable X is exactly equal to a particular value is always zero. This rather curious fact is based on a set of mathematical ideas called measure theory. Intuitively, the probability of finding an individual with exactly some specific characteristic (say, a weight of 2.073192648 kg, or a resistor with a resistance of 10.0372189 Ω) is, well, zero. This is not to say that once you find such an individual, you must be hallucinating. The notion of zero probability (and in fact any probability) relates to "a priori" determination, i.e., before any observation. Once an observation is made, the probability of observing whatever it is you observed is in fact 1 or 100%.

2.2 Mathematical Features of Random Variables

In general, capital letters, like X, will refer to a random variable, whereas lower case letters, like x, will refer to a specific value of the random variable, X. Often, in order to avoid confusing discrete and continuous variables, the symbol $f_X(x)$ will refer to the density function for variable X, evaluated at the value x, and $p_X(x_k)$ to a probability mass function for a discrete variable X evaluated at the value x_k. The notation Pr{ } will refer to the probability that whatever is inside the curly brackets will happen, or be observed. If the symbol "dx" means a very small value for X, and x_k represents a particular value of a discrete random variable, then

$$f_X(x)dx = \Pr\{x - dx \leq X \leq x + dx\}$$

and

$$p_X(x_k) = \Pr\{X = x_k\}$$

There is a particularly important function called the cumulative distribution function (CDF) that is the probability $Pr\{X \leq x\}$, which is usually defined in terms of density or mass functions, namely:

$$F_X(x) = \int_{-\infty}^{x} f_X(\xi)\, d\xi$$

for continuous variables, and

$$F_X(x) = \sum_{x_k \le x} p_X(x_k)$$

for discrete variables.

As mentioned earlier, the functions $f_X(.)$ and $p_X(.)$ generally have parameters, or constants, that dictate something about the particular nature of the shape of the density curve. Table 2.1 provides the parameter lists, and density or mass functions for several common distributions.

In the case of the Binomial and Beta distributions, the symbol "p" was used to denote a parameter (Binomial), or as a value of a random variable (Beta), and not the mass function itself. The function $\Gamma(x)$ is called the gamma function (oddly enough) and has a definition in terms of an integral:

Table 2.1 Some probability density and mass functions

Name	Parameters	Density or Mass function	Expected value	Range of values
Normal	μ, σ	$\frac{1}{\sqrt{2\pi}\sigma}\exp\left(-\frac{1}{2}\left(\frac{x-\mu}{\sigma}\right)^2\right)$	μ	$-\infty < x < +\infty$
Gamma	ν, λ	$\frac{\lambda^\nu}{\Gamma(\nu)}x^{\nu-1}\exp(-\lambda x)$	$\nu\lambda$	$x > 0$
Chi-squared	ν	$\frac{(1/2)^{\frac{\nu}{2}}}{\Gamma\left(\frac{\nu}{2}\right)}x^{\frac{\nu}{2}-1}\exp\left(-\frac{1}{2}x\right)$	$\frac{\nu}{2}$	$x > 0$
Student's t	ν	$\frac{\Gamma\left(\frac{1}{2}(\nu+1)\right)}{\sqrt{\pi\nu}\,\Gamma\left(\frac{1}{2}\nu\right)}\left[1+\frac{x^2}{\nu}\right]^{-\frac{(\nu+1)}{2}}$	0	$-\infty < x < +\infty$
F	ν_1, ν_2	$\frac{\Gamma\left(\frac{\nu_1+\nu_2}{2}\right)}{\Gamma\left(\frac{1}{2}\nu_1\right)\Gamma\left(\frac{1}{2}\nu_2\right)}\frac{x^{\left(\frac{\nu}{2}\right)-1}}{(1+x)^{(\nu_1+\nu_2)/2}}$	$\frac{\nu_2}{\nu_2-2}$	$x > 0$
Poisson	λ	$\frac{\lambda^k e^{-\lambda}}{k!}$	λ	$k = 0, 1, 2, \ldots$
Binomial	n, p	$\binom{n}{k}p^k(1-p)^{n-k}$	np	$k = 0, 1, 2, 3, \ldots n$
Beta	α, β	$\frac{\Gamma(\alpha+\beta)}{\Gamma(\alpha)\Gamma(\beta)}p^{\alpha-1}(1-p)^{\beta-1}$	$\frac{\alpha}{\alpha+\beta}$	$0 < p \le 1$
Truncated normal	μ, σ, a, b $(\alpha = (a-\mu)/\sigma)$, $(\beta = (b-\mu)/\sigma)$	$\frac{\phi(z)}{Z}$ $z = (x-\mu)/\sigma$ $Z = \Phi(\beta) - \Phi(\alpha)$ $\phi(z)$ = *std. normal density*	$u + \frac{\phi(\alpha)-\phi(\beta)}{Z}\sigma$	$a \le x \le b$

$$\Gamma(x)=\int_{0}^{\infty}\xi^{x-1}e^{-x}d\xi$$

Aside from the CDF, there are some other important functions of $f_X(x)$ and $p_X(x_k)$. In particular, there is the expected value, or mean:

$$E[X]=\mu=\begin{cases}\sum_{k}x_k p_X(x_k)\\ \int_{-\infty}^{+\infty}\xi f_X(\xi)d\xi\end{cases}$$

and the variance:

$$V[X]=E\left[(X-\mu)^2\right]=\sigma^2=\begin{cases}\sum_{k}(x_k-\mu)^2 p_X(x_k)\\ \int_{-\infty}^{+\infty}(\xi-\mu)^2 f_X(\xi)d\xi\end{cases}$$

An important property of expectation and variance is the effect of multiplying a random variable by a constant:

$$E[aX]=aE[X]$$
$$V[aX]=a^2V[X]$$

The expected value is also called the first moment about 0, and the variance is called the second moment about the mean. In general, the kth moment about zero is simply:

$$E\left[X^k\right]=\mu^{(k)}=\begin{cases}\sum_{i}x_i^k p_X(x_i)\\ \int_{-\infty}^{+\infty}\xi^k f_X(\xi)d\xi.\end{cases}$$

The kth moment about the mean is:

$$E[X-\mu]^k=\sigma^{(k)}=\begin{cases}\sum_{i}(x_i-\mu)^k p_X(x_i)\\ \int_{-\infty}^{+\infty}(\xi-\mu)^k f_X(\xi)d\xi\end{cases}$$

Commonly the Greek letter μ is used to symbolize the expected value, and σ^2 is used to represent the variance. The variance is never negative (a sum of squared values).

The square root of the variance is called the standard deviation, and has its most important role in random variables having a normal distribution. The expected value has units that are the same as individual measurements or observations. The variance has squared units, so that the standard deviation has the same units as the measurements.

Another parameter that is a measure of variability is the coefficient of variation, or cv. The cv is sometimes referred to as relative standard deviation (rsd). It is the ratio of the standard deviation to the expected value:

$$c = \frac{\sigma}{\mu}$$

Inasmuch as this is a unitless quantity, it is sometimes expressed as a percentage.

2.3 More than One Random Variable

Often we must deal with more than one random variable simultaneously. The density or mass function of one variable might depend on the value of some other variable. Such dependency is referred to as "conditioning." We symbolize the conditional density of X, given another variable, say Y, is equal to a particular value, say y, using the notation:

$$f_{X|Y}\left(x|Y = y\right)$$

Typically, the fact that $Y = y$ will affect the particular values of parameters. Also, we will usually drop the subscript $X|Y$, since the conditional nature of the density is made obvious by the "|" notation.

It is possible that the value of one random variable, say Y, has no effect on the probability distribution of another, X. It turns out that any two random variables have what is called a joint density function. The joint density of X and Y could be defined as:

$$f_{XY}\left(x,y\right)dxdy = \Pr\left\{x - dx \le X \le x + dx, \text{AND } y - dy \le Y \le y + dy\right\}$$

The joint density quantifies the probability that random variable X falls in a given range and at the same time random variable Y falls in some other given range.

It turns out that this joint density can be expressed in terms of conditional densities:

$$f_{XY}\left(x,y\right) = f_{X|Y}\left(x|y\right) f_Y\left(y\right)$$

The marginal density of one variable (say *X*) is the density of *X* without the effect of *Y*, and is computed as:

$$f_X(x) = \int_{-\infty}^{+\infty} f_{XY}(x,y)\,dy$$

In general, two events (an event is a set of possible outcomes or possible values for a random variable), call them *A* and *B*, are independent if their joint probability is equal to the product of their individual probabilities:

$$\Pr\{A \text{ and } B\} = \Pr\{A\}\Pr\{B\}$$

Two random variables are called independent if their joint probability distribution function is the product of their individuals distribution functions:

$$\Pr\{X < x, Y < y\} = \Pr\{X < x\}\Pr\{Y < y\}$$

When *X* and *Y* are independent of each other, and they both have density functions (most cases do), then

$$f_{X|Y}(x|y) = f_X(x)$$

So that

$$f_{XY}(x,y) = f_X(x)\,f_Y(y)$$

In other words, when *X* and *Y* are independent, their joint density is the product of their marginal densities.

Two random variables, *X* and *Y*, may not be independent, and their joint density may not be factorable into two functions, one of *X* and the other of *Y*. There is another function called covariance, which quantifies the degree to which the two variables depend upon each other:

$$\mathrm{cov}[X,Y] = E[(X-\mu_X)(Y-\mu_Y)]$$

Often, the covariance is "normalized," so that its values vary between −1 and + 1. The normalized covariance is called correlation:

$$Corr(X,Y) = \frac{\mathrm{cov}(X,Y)}{\sqrt{V_X[X]V_Y[Y\}}}$$

The values μ_X and μ_Y are referred to as the marginal means of *X* and *Y*, respectively, and Var(*X*) and *Var*(*Y*) are the marginal variances. "Marginal" means with respect to the marginal distributions, e.g.:

$$E_X[X] = \mu_X = \int_{-\infty}^{+\infty} \xi f_X(\xi) d\xi$$

and

$$V_X[X] = \sigma_X^2 = \int_{-\infty}^{+\infty} (\xi - \mu)^2 f_X(\xi) d\xi$$

Correlation is always a number between −1 and +1.

The correlation of any random variable with itself is always +1.

Tow variables are called "uncorrelated" if their correlation is 0. It is not absolutely true that a correlation of 0 between two variables means they are independent. However, most of the time, uncorrelated means independent.

In addition to joint distributions, the expected values and variances of sums and differences of random variables find themselves in many applications. So, if X and Y are two random variables:

$$E[aX \pm bY] = aE[X] \pm bE[Y]$$

If X and Y are independent, then

$$V[aX \pm bY] = a^2 V[X] + b^2 V[Y]$$

where a and b are constants.

While the sign of the operator (±) follows along with the expected values, the variance of the difference is the sum of the variances.

Another set of facts we will use relating to conditional densities or mass functions is based on something called Bayes' Theorem. Briefly, Bayes' Theorem (Gelman et al., 1997) states that if X is a random variable with density f, and Y is a random variable with density g, then

$$g(x|Y = y) = \frac{g(y) f(x|Y = y)}{\int_{-\infty}^{+\infty} f(x|Y = \xi) g(\xi) d\xi}$$

As long as Y is continuous, this particular formula holds even if X is discrete, and f is the mass function of X. If, however, Y is discrete, and g is its mass function, then the integral is replaced with a summation:

$$g(x|Y = y) = \frac{g(y) f(x|Y = y)}{\sum_k f(x|Y = \xi_k) g(\xi_k)}$$

It should be noted that it is possible for a random variable to not actually have a density function associated with it. However, that situation probably never exists in nature, so we will assume the density always exists

2.4 Summary

Probability is the mathematical underlying basis of statistics. Statistics is all about using data to make inferences about populations. Probability provides the experimenter with a way of judging the degree to which those inferences are trustworthy.

2.5 Points to Ponder

1. If X is a random variable with mean 10 and standard deviation 2, what are the mean and standard deviation of $Y = 5X$?
2. Random variables X and Y are independent. $V[X] = 16$ and $V[Y] = 4$. What is the covariance?
3. The density of X is 1/10 for all values from $X = 0$ to $X = 10$. What is the probability (approximately) that X will fall in the interval $(5 - 0.00001, 5 + 0.00001)$?

References

Gelman, A., Carlin, J. B., Stern, H. S., & Rubin, D. B. (1997). *Bayesian data analysis*. Chapman & Hall/CRC.

Meyer, P. L. (1970). *Introduction to probability and statistical applications* (2nd ed.). Addison-Wesley.

Chapter 3
Estimation, Hypothesis Testing, and the Scientific Method

3.1 General Concepts

Statistical problems can be classified into three categories:

1. Estimation.
2. Inference and Decision-Making.
3. Model Building: Discrimination and Prediction.

Estimation is the process of using data to guess at the value of parameters or some feature of the probability distribution assumed to be governing the data-generating process. Probably the most common is estimating the expected value of a distribution. As defined earlier, the expected value of the random variable's distribution is:

$$E[X] = \mu = \begin{cases} \sum_k x_k p_X(x_k) \\ \int_{-\infty}^{+\infty} \xi f_X(\xi) d\xi \end{cases}$$

One of the useful mathematical properties of expected value is that it is a linear operator (Meyer, 1970), namely:

$$E[X_1 + X_2 + \ldots + X_n] = E[X_1] + E[X_2] + \ldots + E[X_n]$$

And

$$E[aX] = aE[X] \text{ when } a \text{ is a non-random constant}$$

S. A. Pardo, *Beyond the T-Test*, https://doi.org/10.1007/978-3-031-84479-9_3

An estimate based on a sample of observations from the data-generating process is:

$$\hat{\mu} = \frac{1}{n}\sum_{i=1}^{n} x_i$$

We use the notation $\hat{\mu}$ instead of a perhaps more well-recognized symbol $\bar{x}$, to emphasize the fact that we are using the data to estimate the expected value. There are many such estimation formulae (called estimators), and many are used in different contexts for different reasons. The main point is that data can be used to estimate parameters or other features of probability distributions. The other point is that, since estimators use data, they themselves are random variables. Thus, if two researchers studying the same population of finches each make independent observations on either two sets (samples) of birds or even on the same sample, but at two different times, and each researcher calculates an average, the two averages most likely won't be the same exactly. Similarly, if two chemists measure the heat of a reaction after a fixed amount of time on each of several samples of reactants, and each chemist computes the average heat, they will probably not obtain exactly the same numerical result.

3.2 Estimation

Formulas often can be derived to provide a way to calculate estimates of parameters. The formulas are referred to as estimators, and the resulting calculation is called an estimate. There are different methods used to derive estimator formulas for various parameters. Perhaps the best known is called the method of maximum likelihood. The idea is that if you have a random sample of measurements (X), you can find values of parameters that maximize something called the likelihood function, which generally depends on assuming the form of the distribution for the data-generating process. Suppose that the values x_1, x_2, …,x_n represent n values sampled from a normally distributed data-generating process, with unknown expected value and variance the density function evaluated at x_i, say, would be given by:

$$f(x_i) = \frac{1}{\sigma\sqrt{2\pi}} e^{-\frac{1}{2}\left(\frac{x_i-\mu}{\sigma}\right)^2}$$

The likelihood function for the sample would be the product of all the valuations of the density function:

$$L(x_1, x_n, \ldots, x_n) = \prod_{i=1}^{n} f(x_i)$$

Of course, this likelihood function cannot be computed without knowing μ and σ. The idea of maximum likelihood is to find values $\hat{\mu}$ and $\hat{\sigma}$ that maximize L. Usually the log of the likelihood function is taken before attempting to solve the maximization. Maximizing the log of L is equivalent to maximizing L, since the log is a monotonic increasing function. The log of a product is the sum of the logs of the factors:

$$\log L = \sum_{i=1}^{n} \log\left(f\left(x_i\right)\right)$$

Maximizing the sum is easier mathematically than maximizing the product.

What is important to note is that first we had to pick a parametric form for the density function of the random variable from which we were sampling, the parameter values are unknown, and our guess for the parameter values is based on a criterion that gives us the "best guess." It turns out that for the normal model, the maximum likelihood estimators for μ and σ^2 are:

$$\hat{\mu} = \frac{1}{n}\sum_{i=1}^{n} x_i$$

and

$$\hat{\sigma}^2 = \frac{1}{n}\sum_{i=1}^{n}\left(x_i - \hat{\mu}\right)^2$$

In general, the kth sample moment about the mean is given by:

$$\hat{\sigma}^{(k)} = \frac{1}{n}\sum_{i=1}^{n}\left(x_i - \hat{\mu}\right)^k$$

Some may notice that the maximum likelihood estimator for σ^2 differs from the formula used in most elementary texts, in that it divides by n and not $n - 1$. Dividing the sum by $n - 1$ to estimate σ^2 gives the formula a property known as unbiasedness. While this is important, in the case of this estimator, the effects are fairly small. Another estimation method is called least squares. Rather than maximize a likelihood function, least squares chooses estimators that minimize an "error" function. A common context for least squares estimation is linear regression. More will be said about least squares. For now, just recognize it as a method for estimating parameters.

Statistical estimates, since they are based on a finite sample of observations or measurements made on individuals taken from some population or data-generating process, have themselves a random variation component. In fact, statistical estimates are themselves random variables, and they have distributions. Their distributions, often called sampling distributions, depend generally upon the distribution of the random variable that was sampled. Inasmuch as a statistical estimate is attempting to help make a guess about a parameter, it would be good to know that the formula used to compute the estimate has a reasonable chance of getting "close" to the actual value of the parameter. One such property has already been described, namely maximum likelihood. Another property that is desirable is called unbiasedness, which was also mentioned earlier. An estimation formula is said to be unbiased if its expected value is equal to the parameter to be estimated. For example, assuming a random sample, $x_1, x_2, \ldots, x_n$, then the expected value of each x_i is the population mean, μ, and:

$$E\left[\hat{\mu}\right] = \frac{1}{n}\sum_{i=1}^{n} E\left[x_i\right] = \frac{1}{n}\sum_{i=1}^{n}\mu = \mu$$

Thus our arithmetic mean estimator for μ is in fact unbiased. Conversely, the maximum likelihood estimate of σ^2 is not unbiased (or, in other words, biased). It turns out that:

$$E\left[\hat{\sigma}^2\right] = \frac{1}{n}\sum_{i=1}^{n} E\left[\left(x_i - \hat{\mu}\right)^2\right] = \frac{n-1}{n}\sigma^2$$

Thus, the maximum likelihood estimator of σ^2 slightly underestimates the variance. The point of the discussion about unbiasedness is that estimation formulae are themselves random variables, and as such we will need to consider their probabilistic characteristics, i.e., their sampling distributions.

3.3 Inference and Hypotheses

In his work of 1892, 'The Grammer of Science" (Pearson, 1892), Karl Pearson, arguably the father of twentieth century statistics, explains that science is a method, and not a body of material. But what is that method? How does it relate to statistics? Briefly stated, the scientific method may be summarized as:

1. Have a question;
2. Formulate an hypothesis about the answer;
3. Perform an experiment to gather data;
4. Analyze the data to test the hypothesis;
5. Conclude that the hypothesis is either correct or not.

Statistics generally starts with questions about some "parameters"; numbers that provide the answer to the question. Inference is about making "a priori" guesses about parameter values, and then using data to decide if you were correct. Suppose, for example, you guessed that the average duration of a courtship display was 30 s. How would you decide whether to believe your guess, or not? First you would gather data, by timing courtship displays of some number of individuals, say n. Then you would probably compute the maximum likelihood estimates of mean and variance. Suppose the estimate of the mean was 31.5 s, and the standard deviation (square root of variance) estimate was 3 s. OK, so it wasn't 30. Were you wrong? The question becomes one of how much variation there might be if the experiment were repeated. The idea of statistical inference is to make a decision about what to believe, and not what actually is the truth. Our decision has risk associated with it, namely the risk (or probability) of saying our guess is wrong when in fact it is correct, and the risk of saying our guess is correct when in fact it is not. There is a formalism for expressing the notions of inference. There are two competing "hypotheses," or guesses about the parameter or parameters of interest. One is called the "null" hypothesis, symbolized as "H_0." The logical negation of the null hypothesis is called, not surprisingly, the alternate hypothesis, and is often symbolized as "H_1". So, in the example of the courting display question, we might have:

H_0: $\mu = 30$.
H_1: $\mu \neq 30$.

The process of formulating hypotheses about parameters, and then using data to decide which of the two (mutually exclusive) hypotheses to believe is referred to as "hypothesis testing" (Hoel, 1971).

The error of deciding that H_0 is false when in fact it was true is called Type I error. The error of believing H_0 is true when it is not is called Type II error. The next thing required is a rule, based on data, that lets the decision-maker decide whether to believe H_0 or H_1. Since data are subject to variation, the rule is necessarily probabilistic. It turns out that, conveniently, the calculation:

$$t = \frac{\hat{\mu} - 30}{\hat{\sigma}/\sqrt{n}}$$

has a known probability (sampling) distribution, the familiar Student's t, provided that the null hypothesis is actually correct (i.e., that $\mu = 30$). This formula is known as a test statistic, because it is the quantity we will use to decide whether to believe (accept) the null hypothesis, or disbelieve it (reject) The use of this statistic to test hypotheses is referred to as a t-test. It is important to note that the t statistic has Student's t distribution under the assumption that the data were sampled from a population having a normal distribution. However, it is true that as the sample size gets "large," sample means will have a normal distribution with expected value:

$$E\left[\hat{\mu}\right] = \mu$$

and variance:

$$V\left[\hat{\mu}\right] = \frac{\sigma^2}{n}$$

The convergence of the sample mean's distribution to normal is called the "central limit theorem" (Hoel, ref.cit.). In fact, it is one of several forms of the central limit theorem, but maybe the most recognizable. Using these facts allows for finding closed-form probabilistic limits on the test statistic *t*.

A common feature of all inference is determining the distribution of the test statistic if H_0 were actually true. The probability of making Type I error is symbolized with the letter α. The probability of Type II error is traditionally symbolized with the letter β. We can find a range of values that *t* would fall in between with probability $1 - \alpha$, given that H_0 is true, even before we gathered any data. In fact, the range of possible values only depends on the sample size, *n*, and the desired probability content of the range. If, for example, the sample size was $n = 10$, and we wanted the probability content to be $100(1 - \alpha)\% = 95\%$, then the range of values for *t* we would expect if the null hypothesis was correct would be approximately ± 2.228. The range $(t \leq -2.228,\ t \geq +2.228)$ is called the *critical region* of size α. If the value of the test statistic falls in the critical region, we say that the test statistics is significant at the $100\alpha\%$ level, and we REJECT the null hypothesis is favor of the alternative. The particular region for this example is partially based on the presumption that we computed the maximum likelihood estimate for standard deviation. If after getting data, we computed the value of *t* using the formula above, and its value fell within the range $\pm$ 2.228, then we would continue to believe the null hypothesis, because there is a fairly "high" (95%) chance of *t* falling inside this range if H_0 is correct. Conversely, there is a relatively "low" chance that *t* would fall outside the "critical" range if H_0 was correct. Unfortunately, we cannot make the same statement about the alternative hypothesis, H_1, since there are an infinite number of possible values (anything other than 30) that would make it correct. Thus, it is easier to fix the chance of making the mistake of deciding that H_0 is false when in fact it is true. Once this risk is decided upon, the decision rule for either believing H_0 or not believing it is fairly easy to compute, provided we know something about the distribution of the test statistic, given that the null hypothesis is true.

Another way of determining a rule for rejecting or accepting the null hypothesis is to *compute a probability* of observing the data you got IF the null hypothesis was actually correct. This probability is usually referred to as a "*p*-value." Thus, in our example, if in fact $\mu = 30$, then the test statistic:

$$t = \frac{\hat{\mu} - 30}{\hat{\sigma} / \sqrt{n}}$$

has a Student's *t* distribution with degrees-of-freedom parameter equal to *n* (since we used the maximum likelihood estimate of σ). Suppose we had data that yielded a sample estimate of μ, say:

$$\hat{\mu} = \frac{1}{n}\sum_{i=1}^{n} x_i = 31.5$$

and an estimate of σ^2:

$$\hat{\sigma}^2 = \frac{1}{n}\sum_{i=1}^{n}\left(x_i - \hat{\mu}\right)^2 = 9$$

If $n = 25$, then the sample test statistic would be:

$$t = \frac{31.5 - 30}{3/\sqrt{25}} \approx 2.50$$

Since the alternative hypothesis is $\mu \neq 30$, we compute the probability that the test statistic would be outside the range (−2.50, +2.50). To compute this, we can use the R function *pt*():

```
pt(q=-2.5,df=25,lower.tail=TRUE) ≈ 0.01934
and
pt(q=+2.5,df=25,lower.tail=FALSE) ≈ 0.01934
```

The "two-sided" *p*-value is 0.01934 + 0.01934 = 0.03868.

Since Student's *t* distribution is symmetric about 0, the probability for the "lower tail" of $-t$ is equal to the probability for the "upper tail" of $+t$.

If our threshold of *p*-values is $\alpha = 0.05$, then since 0.03868 is less than 0.05, we will no longer believe that the null hypothesis is correct, and reject it. With a sample size of $n = 25$, and $1—\alpha = 0.95$, then the critical region is $t \leq -2.06$, $t \geq +2.06$. Since $t = 2.50 > +2.06$, we would reject the null hypothesis. Regardless of whether you determine a critical region of size α, or choose α to be a threshold for p-values, the conclusions would be identical.

Another concept that is somewhere between estimation and inference is called confidence interval building. The confidence interval again employs that risk level, α, but in a slightly different manner. Until now, estimates were only a single number, such as $\hat{\mu}$. Such an estimate is referred to as a "point" estimate. Suppose we wanted to know the values of the parameters that would correspond to the limits of the critical range for the test statistic. Using the previous example, let:

$$t_{\text{low}} = -2.06 = \frac{\hat{\mu} - \mu_{\text{low}}}{\hat{\sigma}/\sqrt{n}}$$

and

$$t_{\text{high}} = -2.06 = \frac{\hat{\mu} - \mu_{\text{high}}}{\hat{\sigma}/\sqrt{n}}$$

Solving for μ_{low} and μ_{high} gives:

$$\mu_{\text{low}} = \hat{\mu} - 2.06\frac{\hat{\sigma}}{\sqrt{n}}$$

and

$$\mu_{\text{high}} = \hat{\mu} + 2.06\frac{\hat{\sigma}}{\sqrt{n}}$$

The range of values (μ_{low}, μ_{high}) is called the 100(1 − α)% "confidence interval" for parameter μ. It can be thought of as a feasible range for the unknown values of μ. That is, we are not certain about the actual value of μ, but we are nearly certain (100(1-α)% certain) that it lies somewhere in the interval (μ_{low}, μ_{high}). Technically, we say that the interval (μ_{low}, μ_{high}) has a 100(1-α)% chance of covering, or containing, the parameter value. In any case, we are nearly certain that this interval contains the actual population mean. So, in our example with point estimates $\hat{\mu} = 31.5$, $\hat{\sigma} = 3$, and $n = 25$, the 95% confidence interval would be:

$$\mu_{\text{low}} = 31.5 - 2.06\frac{3}{\sqrt{25}} \approx 30.26$$

$$\mu_{\text{high}} = 31.5 + 2.06\frac{3}{\sqrt{25}} \approx 32.74$$

Since the hypothetical value for μ, namely 30, is not contained in the confidence interval (30.26, 32.74), we do not believe that 30 is a feasible value for μ.

When the null hypothesis is rejected, we say that the difference between our estimate of the parameter and the null value is *statistically significant at the 100α% level*. Another way of stating the same thing is that if we reject the null hypothesis, we would believe that the results of our analyses are repeatable.

3.4 Models

Model building is a special application of estimation, and it usually has some inference associated with it. The idea is to postulate some mathematical relationship between some variables, some random and some without any random component. Then we estimate the values of model parameters. Finally, we test to see if we should believe that the form of the model we postulated was reasonable. Thus modeling has elements of both estimation and inference. Models can be predictive or discriminatory/classificatory. A simple example of a predictive model would be a simple linear regression (Draper & Smith, 1981). Suppose there is a continuously-valued random variable, Y, and another continuously values non-random variable, X. Y could be things such as response time, elapsed time, distance traveled, or other random variables that can be expressed as a decimal number. In this simple case, we are assuming the X variable is not random. In other words, X is something whose value would have no random variation, and whose value is known perfectly without error. Y is referred to as the response variable, and X is the predictor or regressor variable. The general form of the model is:

$$Y = \beta_0 + \beta_1 X + \varepsilon$$

The coefficients β_0 and β_1 are unknown, and need to be estimated. The variable ε represents random "noise," indicating that the value of Y is on the average a linear function of X, but the actual observed values may have some perturbations, or noise, or sometimes called error associated with them.

Once the values of the parameters are estimated, a predicted value of Y can be computed for a given value of X. We would not usually consider X to have random noise associated with it. That is, when we get a value for X, we are (mostly) certain that the value would not vary if we measured or observed it a second time under exactly the same conditions. Rather, we suppose that given the value of X, we can predict on the average what Y would be, with the understanding that Y might vary from this average.

Another closely related type of model is also linear, but is classificatory or discriminatory. The "X" variables are not continuous, but are discrete categories. The goal is to determine if particular groupings of individuals actually discriminate between individuals. In other words, we want to know if individuals in different groups actually differ from each other with respect to Y. Perhaps the simplest example is the one-way Analysis of Variance (ANOVA). In this case, the single X variable is a set of discrete categories, possibly more than 2, and Y is the continuous random variable response. The question is not to find a "prediction" of Y for a given value of X, per se. Rather, the question is to estimate the difference in the average Y between the different categories. In the case of ANOVA, often the inferential part of modeling is of greater interest, namely whether the difference in average values of Y between the different groups of the X categories is in fact repeatable.

There are certainly more types of both predictive and classificatory modeling. The key notion here is that data can be used to create these sorts of models, through a combination of estimation and inference.

This is the classical parametric methodology for statistical inference. There is another set of methods, sometimes called nonparametric or distribution-free, neither of which term is strictly true. The idea is that the distribution of test statistics should not depend on the distribution of the data-generating process. The basic idea is still the same; you formulate a test statistic, you determine the "critical range" or "critical value" based on α, you get some data, and then you compute the test statistic to decide if you should accept or reject the null hypothesis.

A special set of nonparametric techniques is sometimes referred to as resampling methods. This book will in fact emphasize resampling methods where appropriate. The resampling techniques will generally fall into the bootstrap estimation process or the permutation hypothesis testing process. Both of these methods are computer-based, but given modern computing software such as R, they are fairly easy to perform.

3.5 Bayes' Theorem

Bayesian statistics is an alternate view of parameters, not as particular values to estimate or about which to make a guess about their true values, but treating them as if they themselves are random variables. Like the classic "frequentist" approach, Bayesian methods employ a likelihood function. However, these methods incorporate prior information about the parameters of interest. "Prior" to making observations, the analyst posits a distribution of the parameters of interest. The "prior" distribution expresses the knowledge about the parameter prior to performing the "next" experiment. So, for example, perhaps the mean response time to a stimulus is guessed to be most likely 10 s, but it could be as fast as 5 s and as delayed as 15 s. Rather than simply hypothesizing that the mean is exactly 10 s, the Bayesian method is to postulate a distribution that expresses the current level of knowledge and uncertainty in the parameter. Then, once data are gathered, Bayes' theorem is used to combine the prior distribution with the likelihood function, to update the prior knowledge. The updated distribution for the parameter is called the posterior distribution. So, if $f_{\text{old}}(\tilde{\mu})$ represents the prior density function for the parameter $\tilde{\mu}$, and $L[x_1,|x_2,|\ldots,|x_n,|\tilde{\mu}]$ the likelihood function for the sample, given a particular value of $\tilde{\mu}$, then the updated density function (called the posterior density) is:

$$f_{\text{new}}(\tilde{\mu}|,x_1|,x_2|,\ldots|,x_n) = \frac{f_{\text{old}}(\tilde{\mu})L[x_1,|x_2,|\ldots,|x_n,|\tilde{\mu}]}{\int_{-\infty}^{+\infty} f_{\text{old}}(\xi)L[x_1,|x_2,|\ldots,|x_n,|\xi]\,d\xi}$$

3.6 Points to Ponder

1. A random sample of eight observations from a normal distribution is:
   1.0, 2.1, 3.7, 4.8, 3.6, 7.1, 2.2, 4.4.
   What are the maximum likelihood estimates of mean and variance?
2. Suppose it is uncertain whether for some process, the probability of "success," p, is 90% or 99%. Suppose further that, based on various experiences together with some theoretical knowledge, we believe that there is a 60% chance that $p = 0.90$ (90%) and a 40% chance that $p = 0.99$ (99%). We do an experiment in which we run the process $n = 100$ times, and the number of times it results in a success is $X = 89$. Would you change your thoughts about whether $p = 0.90$ or $p = 0.99$? If so, how would you go about making that change? Hint: Bayes' Theorem might help.
3. In an experiment with n = 15 units, the sample mean = 10 and the sample standard deviation = 2. Can you compute a p-value based on the null hypothesis that the population mean is actually 12?

References

Draper, N. R., & Smith, H. (1981). *Applied regression analysis* (2nd ed.). John Wiley and Sons.
Hoel, P. (1971). *Introduction to mathematical statistics*. John Wiley and Sons.
Meyer, P. (1970). *Introductory Probability and Statistical Applications* (2nd ed.). Addison-Wesley Publishing Company, Reading.
Pearson, K. (1892). *The grammer of science*. Walter Scott.

Chapter 4
Binary Random Variables and Acceptance Sampling Plans

4.1 Bernoulli and Binomial Random Variables

A very important class of discrete random variables are those where there are exactly two possible values. The values are generally chosen to be 1 or 0, representing "good" and "not good," or "desirable" and "undesirable" or "success" and "failure" outcomes. For this discussion, the desirable outcome will be associated or mapped to the number 1, and the undesirable outcome with 0.

So mathematically, we can represent this binary variable as:

$$X = \begin{cases} 1\,\text{if desireable} \\ \text{or} \\ 0\,\text{if undesireable} \end{cases}$$

Such a variable is referred to as a Bernoulli variable. Its probability mass function is fairly simple, namely $\Pr\{X = 1\} = p$ and $\Pr\{X = 0\} = 1—p$. So p is a number between 0 and 1. The iconic example is the "fair coin flip," where $p = 0.5$.

The Bernoulli variables by themselves are not generally very interesting. They get more interesting when you have a sample of them. We will assume a random sample, namely that each Bernoulli outcome is independent of the others, and that that value of p stays the same for each of the values in the sample.

Suppose we have a sample of n Bernoulli variables, the sum of those values will equal the number of times the Bernoulli outcome was desirable. So if $x_1, x_2, \ldots, x_n$ represent the n Bernoulli results then the number of those sample results that equal 1 is:

Supplementary Information The online version contains supplementary material available at https://doi.org/10.1007/978-3-031-84479-9_4. The videos can be accessed individually by clicking the DOI link in the accompanying figure caption or by scanning this link with the SN More Media App.

S. A. Pardo, *Beyond the T-Test*, https://doi.org/10.1007/978-3-031-84479-9_4

$$S = \sum_{i=1}^{n} x_i$$

This sum, *S*, is another random variable whose distribution is called Binomial. Note that before we do the sampling, we do not know the value of *S*. That is, S could be 0 (all the $x_1, x_2, \ldots, x_n$ are 0) or it could be *n* (all the $x_1, x_2, \ldots, x_n$ are 1) or it could be something in between. The number *p* is a parameter, and its value may be unknown. If, however, it was known, the probability that S was equal to any particular value, say, is:

$$\Pr\{S = s|,n|,p\} = \binom{n}{s} p^s (1-p)^{n-s}$$

The symbol $\binom{n}{s}$ is often pronounced "*n* choose *s*" and can be expressed in terms of factorials. The symbol *x*! means $x*(x-1)*(x-2)*\ldots*1$ and is called "*x* factorial." Thus:

$$\binom{n}{s} = \frac{n!}{s!(n-s)!}$$

We will go between using "*" and simply writing two quantities next to each other with no space in between to mean multiplication.

4.2 Binary Variables and Hypothesis Testing

Often there is some hypothesis about the value of *p*. For example, we might hypothesize that the probability *p* (sometimes referred to as the probability of success for each Bernoulli variable) is close to 1, say 0.99 (i.e., 99%). If we know how many Bernoullis we intend to sample (*n*), then we could compute the probability that $S = 0, 1, 2, \ldots, n$.

Rather that interest specifically in the probability mass function values, the interest may be in the probability that *S* will equal to at least some particular value, for example, approximately 95% of *n*. Remember that *S* is discrete, and must equal to an integer. *S* cannot equal 50.7, for example. Thus, if $n = 92$, then 95% of *n* is 87.4. *S* could be 87 or 88. If we are hoping that *S* will be at least 95% of *n*, then if $n = 92$, our target value would be 88.

The requirement for the number of "successes" is referred to as the critical value. We can determine the risk associated with that critical value based on the sample size and the null hypothesis.

Suppose we really wanted the probability of a "success" (i.e., the population parameter) to be 0.95 (95%). If our critical value is 88 out of 92 (i.e., 88 is just a bit over 95% of $n = 92$), i.e., "PASSING," what is the chance of achieving this critical value IF the population parameter is actually 0.95? The answer comes from the cumulative distribution function (CDF) of the Binomial random variable. It is:

$$\Pr\{S \geq 88 | ,n = 92 | ,p = 0.95\} = 1 - \sum_{k=0}^{87} \binom{92}{k} (0.95)^k (1 - 0.95)^{92-k} \approx 0.5104$$

So if $p = 0.95$ with $n = 92$, there is only about a 51.04% chance of PASSING, or getting at least 88 "successes." A sample size together with a critical value is called a sampling plan. If achieving the critical value means that somehow we find the population to be acceptable (i.e., PASS), the plan is called an Acceptance Sampling Plan (Schilling, 1982). So, it looks like a critical value of 88 with a sample size of $n = 92$ is not realistic if we really hope that $p = 0.95$. Suppose we changed the critical value to 83. Then:

$$\Pr\{S \geq 83 | ,n = 92 | ,p = 0.95\} = 1 - \sum_{k=0}^{82} \binom{92}{k} (0.95)^k (1 - 0.95)^{92-k} \approx 0.9833.$$

So it looks like 83 is a more favorable critical value. But note that while 88 is about 95.65% of 92, 83 is only about 90.22% of 92. Conversely, if we have our heart set on a critical value of at least 95% of 92, then we would need p to be at least 0.979 in order to have a higher chance of PASSING:

$$\Pr\{S \geq 88 | ,n = 92 | ,p = 0.979\} = 1 - \sum_{k=0}^{87} \binom{92}{k} (0.979)^k (1 - 0.979)^{92-k} \approx 0.9551$$

4.3 Power

We have established that if we decide that at least 95% of the 92 items we sampled must be "successes" in order to be satisfied that the population is acceptable, the population parameter must be at least 0.979 in order to have at least a 95% chance of PASSing. We can also ask the question: How bad must the population be in order to have a very low chance of achieving our critical value? The answer is about $p = 0.9035$. That is, if the population probability of a "success" is only about 0.9035, there is a fairly low chance that in our sample of 92 items, at least 88 will be "successes." How low?

$$\Pr\{S \geq 88 | ,n = 92 | ,p = 0.9035\} = 1 - \sum_{k=0}^{87} \binom{92}{k} (0.9035)^k (1 - 0.9035)^{92-k} \approx 0.0507$$

Since we do not know that actual value of p, we can only hypothesize about it. Suppose our hypotheses are:

$$H_0 : p < 0.95$$

And the alternative, its logical negation:

$$H_1 : p \geq 0.95$$

In the hypothesis testing paradigm, there are generally two hypotheses, and only one of them could be true. So how do we know which is true? Well, we never actually know. In this formulation, if we achieve our critical value for the number of successes, we will REJECT H_0 and ACCEPT H_1. Is it possible that we could get at least 83 successes out of $n = 92$ items even if H_0 is true? Well, suppose that p actually equals 0.9499. That would make H_0 true, since $0.9499 < 0.9500$. So what are the chances of observing at least 83 successes out of $n = 92$, if $p = 0.9499$? You guessed it:

$$\Pr\{S \geq 83|,n = 92|,p = 0.9499\} = 1 - \sum_{k=0}^{82} \binom{92}{k} (0.9499)^k (1 - 0.9499)^{92-k} \approx 0.9831$$

Wow. It is fairly likely that we would observe at least 83 successes out of $n = 92$ EVEN if H_0 is false. We have discovered that there is a "degree" of falseness and a degree of trueness here. Suppose $p = 0.85$. The chance of observing at least 83 successes out of $n = 92$ is:

$$\Pr\{S \geq 83|,n = 92|,p = 0.85\} = 1 - \sum_{k=0}^{82} \binom{92}{k} (0.85)^k (1 - 0.85)^{92-k} \approx 0.1001$$

Power is the probability of rejecting the null hypothesis, and that probability depends on how far from the truth the null hypothesis is. There is a continuum of chances of achieving the critical value for all possible values of p. Such a continuum is called a power curve (sometimes an operating characteristic curve). Fig. 4.1 shows the power curve with $n = 92$, critical value = 83.

Notice that the curve crosses the 5% probability of "PASSING" at $p = 0.849$, and it crosses the 95% probability of "PASSING" at $p = 0.949$. Due to the fact that we cannot have, for example, 83.2 successes, the curve will not cross the 95% chance of PASSING at exactly $p = 0.95$.

One other thing to remember is that power is the probability of rejecting the null hypothesis (H_0) under different possible values of the population parameter(s) of interest.

You may have determined that $n = 92$ is a relatively unusual sample size. For Binomial variables, there are two parameters, namely n and p. The question of what n to choose is perhaps the most common questions asked of statisticians. The answer lies in the computation of power.

Power for a Binomial hypothesis test is a function of n, the critical value (we will call it s_c) and p. So this is one equation with three unknowns. Two of those unknowns must be set to some values in order to find the value of the third. The solution is therefore not unique.

Consider cases with $n = 50$ or $n = 100$. If we set the critical value of be 90% of n, we will not have the same power for a given value of p. With $n = 50$, $s_c = 45$, and for $n = 100$, $s_c = 90$. At $p = 0.95$, the power with $n = 50$ is approximately 0.9622. With $n = 100$, $s_c = 90$, the power at $p = 0.95$ is approximately 0.9885. In both cases, the critical values were $0.9n$, or 90% of the sample size. But the power at $p = 0.95$ was different.

(Code 4.1)

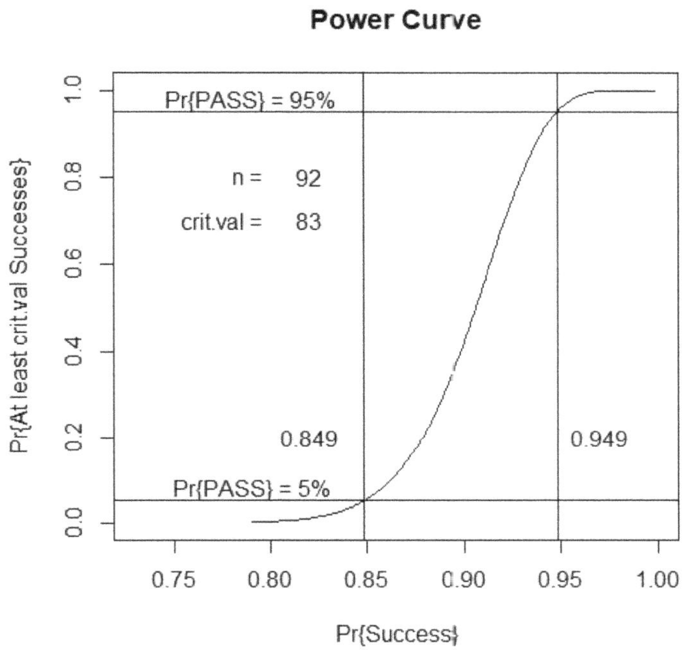

Fig. 4.1 Power curve for a binomial hypothesis test

Table 4.1 AQL and RQL for some sampling plans

(Code 4.1)

Binomial Sampling Plans

Sample Size	Critical Value	AQL	RQL	P.crit	AQL-RQL
50	45	0.9600	0.8260	90.00%	13.40%
92	83	0.9485	0.8485	90.22%	10.00%
100	90	0.9445	0.8480	90.00%	9.65%

AQL = Acceptable Quality Level
RQL = Rejectable Quality Level

Note that as the sample size increases, the difference between AQL and RQL decreases
The power curves for these plans are shown in Fig. 4.2

There are two special values of p; one is called acceptable quality level (AQL) and the other is called rejectable quality level (RQL). AQL is the lowest value of p for which we would say that the population's probability of success is acceptable. RQL is the highest value of p for which we would say that the population's probability of success is unacceptable (i.e., rejectable).

(Code 4.1)

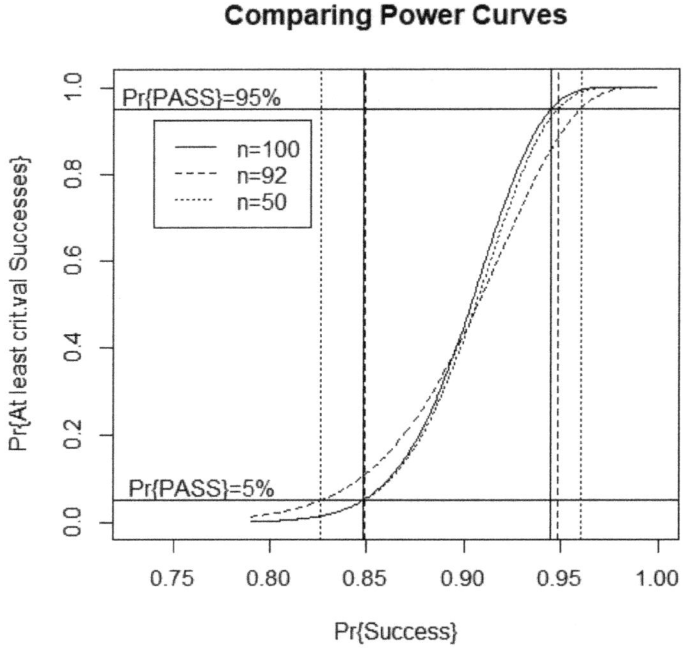

Fig. 4.2 Power curves for three sampling plans

Commonly, acceptance sampling plans are designed so that Pr{*PASS*|*p* = AQL} ≈ 0.95 and Pr{*PASS*|*p* = RQL} ≈ 0.05. Table 4.1 provides the values of plans with n = 100, 92, and 50 for critical values approximately equal to 90% of the sample size.

The power curve is sometimes referred to as the risk curve for the sampling plan. It shows the "risk" or PASSing at various possible values of p.

The power curve can also be constructed by the converse of the sampling plans described here. For example, with n = 92, instead of looking for at least 83 "successes," we could be counting "failures," so that the critical value is 92–83 = 9. That is, we could say that we FAIL if there are more than nine failures observed out of n. The power curve for this "view" of the plan is shown in Fig. 4.3

There are several ways to plot a power curve, depending on what is put on the x-axis (p = Pr{success} or 1—p = Pr{Failure}) and what is on the y-axis (Pr{PASS} or Pr{FAIL}). AQL is an hypothetical population parameter, whose interpretation depends on whether the critical value is the minimum number of "successes" out of n or the maximum number of "failures" out of n required to PASS.

The sampling units are not necessarily discrete "items." They could be attempts to perform some task, where the outcome is either "success" or "failure," or they could be aliquots of some substance from a "pool." The important assumptions are that the outcome for each sampling unit is independent of the others and that the probability of a success for each unit is the same.

(Code 4.1)

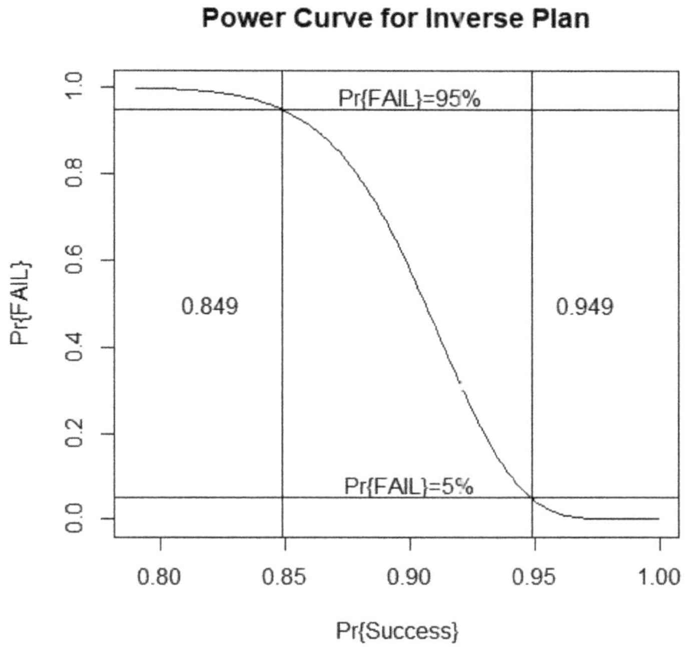

Fig. 4.3 Power curve for sampling plan with $n = 92$, critical value (of failures) = 9

4.4 Estimation

To estimate p, simply average the values of n Bernoulli variables:

$$\hat{p} = \frac{1}{n}\sum_{i=1}^{n} x_i$$

The carat or hat over the p indicates that this is a sample estimate of the population parameter. Since the estimate is based on data, it is not necessarily equal to the parameter value. In fact, if we obtained a different sample from the same population with the same sample size, we are most likely going to obtain a different value of the estimate. Since we generally do not have the liberty of sampling an arbitrarily large sample size, nor can we generally sample with size n over and over, obtaining a collection of estimates, we must rely on a method for assessing the error in the estimate. Such methods are referred to as confidence interval construction. Those methods will be discussed in another chapter.

4.5 Finite Populations and Binary Variables

Sometimes a population is finite but either the sampling process is onerous or expensive, or despite the finite nature of the population it is still too large to observe every unit in the population. In such cases, it may be that the random variable X = number of "successes" out of a sample of size n has what is called a "hypergeometric" distribution. This distribution has a mass function:

$$\Pr\{X = x|N, n, p\} = \frac{\binom{pN}{x}\binom{N-pN}{n-x}}{\binom{N}{pN}}$$

N = population size.
n = sample size,
p = probability of "success."

Note that pN is the population number of successes. If pN is not an integer, it should be rounded to the nearest integer (it is not possible to have a fractional number of successes).

The hypotheses would be stated in the same fashion as it was for the binomial case.

$$H_0 : p < p_0$$

$$H_1 : p \geq p_0$$

The power formula would be different:

$$\Pr\{X \geq x_c|N, n, p\} = 1 - \sum_{k=0}^{x_c-1} \frac{\binom{pN}{k}\binom{N-pN}{n-k}}{\binom{N}{pN}}$$

Figure 4.4 shows a power curve for a situation where the population size is $N = 1000$, the sample size is $n = 100$, and $p_0 = 0.95$. With $x_c = 92$, the power to reject H0 is approximately 95% at $p = p_0 = 0.95$.

The binomial distribution is a good approximation to the hypergeometric distribution, when the sample size is not greater than 20–25% of the population size. As am example, consider the sample size of $n = 100$, with the critical value of $x_c = 92$. However, instead of the population size of $N = 1000$, suppose the population size was $N = 475$. Fig. 4.5 shows the power curves based on the hypergeometric and binomial distributions.

(Code 4.2)

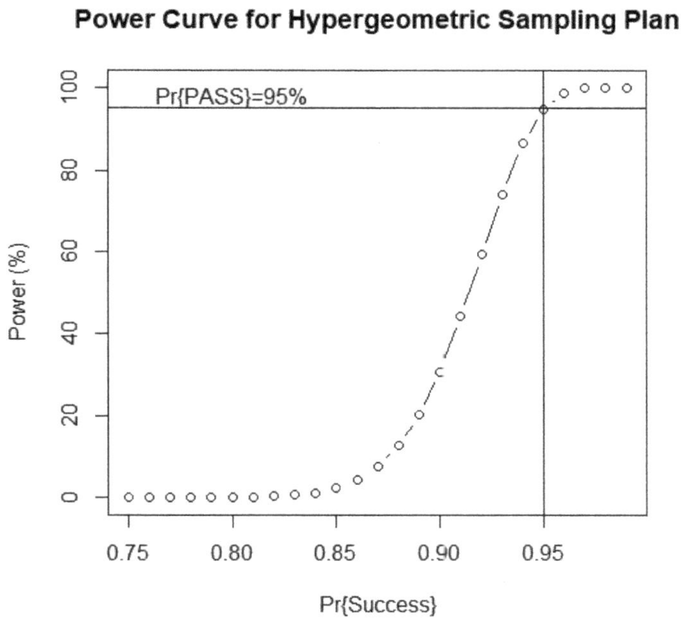

Fig. 4.4 Power curve—hypergeometric distribution

(Code 4.2)

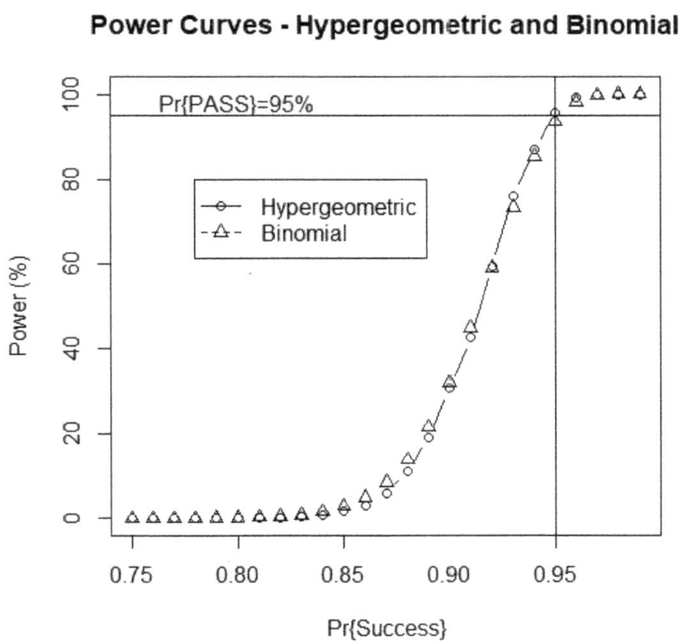

Fig. 4.5 Power cures for $n = 100$, $xc = 92$, $N = 475$ with binomial approximation

4.6 Points to Ponder

1. A product is labeled as being 99% "reliable." In a test of $n = 100$ units, the quality engineer has decided that if 98 out of 100 are reliable, the test is a "PASS." Can you express an hypothesis test for this "reliability" test? What is the critical value?
2. What is AQL? RQL? Which do you think a regulatory agency would find to be more important in assessing the utility of an acceptance test?

Reference

Schilling, E. G. (1982). *Acceptance sampling in quality control.* Marcel Dekker,, ASQC Press.

Chapter 5
Continuous Variables, the Normal Distribution, and the Central Limit Theorem

5.1 The Normal Distribution

Continuously-valued random variables also have probability distributions, and those distributions may be described by some set of parameters. The most common of those distributions is the normal or Gaussian distribution. The normal family of random variables have two parameters, mean and standard deviation, symbolized by the Greek letters μ and σ, respectively. So if X represents the random variable, recall that the expected value is the mean:

$$E[X] = \mu$$

The standard deviation is:

$$\sqrt{E\left[(X-\mu)^2\right]} = \sigma$$

The values of these two parameters completely determine the shape of the density function for a normally distributed random variable. As mentioned earlier, the density function for the normal family is (Johnson et al., 1995):

$$f(x) = \frac{1}{\sigma\sqrt{2\pi}} e^{-\frac{1}{2}\left(\frac{x-\mu}{\sigma}\right)^2}$$

Supplementary Information The online version contains supplementary material available at https://doi.org/10.1007/978-3-031-84479-9_5. The videos can be accessed individually by clicking the DOI link in the accompanying figure caption or by scanning this link with the SN More Media App.

S. A. Pardo, *Beyond the T-Test*, https://doi.org/10.1007/978-3-031-84479-9_5

There is a useful approximation for the percentage of values of X under the normal density curve:

$$X = \mu \pm 1\sigma \rightarrow 68.27\%$$

$$X = \mu \pm 2\sigma \rightarrow 95.45\%$$

$$X = \mu \pm 3\sigma \rightarrow 99.73\%$$

So, for ANY normally distributed variable, approximately 68.27% of the population values will fall within $\mu \pm 1\sigma$, approximately 95.45% will fall within $\mu \pm 2\sigma$, and approximately 99.73% will fall within $\mu \pm 3\sigma$.

Figure 5.1 illustrates a normal density curve with $\mu = 100$ and $\sigma = 10$.

So, in this particular case, approximately 68.27% of the values will be between 90 and 100, approximately 95.45% between 80 and 120, and approximately 99.73% between 70 and 130.

Another feature of normal variables is standardization. If X is a normally distributed variable with mean μ and standard deviation σ, then the variable:

$$Z = \frac{X - \mu}{\sigma}$$

also has a normal distribution with mean $\mu = 0$ and standard deviation $\sigma = 1$. Oddly, Z is called the "Standard Normal" variable, and its values are called "z-scores."

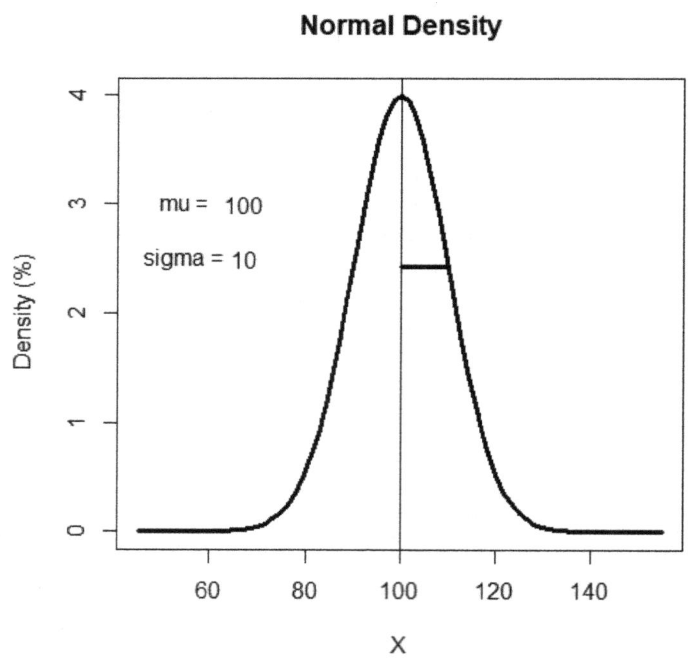

Fig. 5.1 A normal density curve

Thus, about 99.73% of the values of the Standard Normal distribution will fall within the interval (−3,+3), about 95.45% will fall within the interval (−2,+2), and about 68.27% will fall within (−1,+1).

We also use the notation:

$$z = \Phi^{-1}(p)$$

and

$$\Phi(z) = \Pr\{Z \le z\} = \frac{1}{\sqrt{2\pi}} \int_{-\infty}^{z} e^{\frac{1}{2}\xi^2} d\xi$$

5.2 The Central Limit Theorem

The reasons for calling this "bell-shaped" density "normal" are uncertain. Nevertheless, there is nothing particularly normal about it, except for a phenomenon referred to as "The Central Limit Theorem" (Meyer, 1970), which is really a constellation of several theorems. The most common of these is that as you choose larger and larger sample sizes, the sample (arithmetic) means tend to have a normal distribution, whose mean is the mean of the population from which you are sampling and whose standard deviation is the standard deviation of the sampled population divided by the square root of the sample size. So, suppose you samples 100 items from a population whose density was not symmetric about its mean, μ. You calculate the arithmetic mean of those 100 items (i.e., the mean of the random variable of interest). Then you throw those items back into the population, and sample again, with $n = 100$. You continue to perform this process until you have many arithmetic means. Well, the distribution of those sample means will tend to look normal.

Clearly, as you increase the sample size, n, the standard deviation of the arithmetic means gets smaller and smaller. That is, the standard deviation of the arithmetic means, called the standard error, is:

$$\text{SE} = \frac{\sigma}{\sqrt{n}}$$

So, since n is in the denominator, as you make it larger, the quantity SE gets smaller and smaller. Inasmuch as the standard deviation determines the width of the density function, as SE gets smaller, the density of the arithmetic means gets thinner and thinner. So, for example, if $\sigma = 10$ and $n = 100$, the standard error of the arithmetic mean is:

$$\text{SE} = \frac{\sigma}{\sqrt{n}} = \frac{10}{\sqrt{100}} = 1$$

That implies that if the arithmetic means are getting close to having a normal distribution, and if $\mu = 100$, 99.73% of them would fall within $100 \pm 3 = (97,103)$. What is the sample size was $n = 1000$? Now the standard error is only:

$$\mathrm{SE} = \frac{\sigma}{\sqrt{n}} = \frac{10}{\sqrt{1000}} \approx 0.3162$$

So with $n = 1000$, 99.73% of the arithmetic means would be within 100 ± 0.316 3 = (99.0513,100.9487).

So, there are two concepts to note. First, no matter what sort of distribution the population has, the distribution of sample means will get more and more like a normal distribution, centered on the mean of the population from which the sample values were drawn, as you increase the sample size. Second, increasing the sample size narrows down the possible values for the arithmetic mean. In fact, the larger the sample size, the closer the sample arithmetic mean will get to the population mean, μ. Maybe that is why they call the standard deviation of arithmetic means the standard error; the error gets smaller as the sample size is increased.

5.3 Other Continuous Random Variables

While the normal density is symmetric about its mean, μ, there are other random variables for which that is not true. Another important family of continuous random variables is the gamma family. Gamma variables have two parameters, often referred to as shape (we will symbolize is with the Greek letter ν) and scale (symbolized with the Greek letter λ). The density function (Johnson et al., 1995, ref. cit.) is:

$$f(x) = \frac{\lambda^{\nu}}{\Gamma(\nu)} x^{\nu-1} \exp(-\lambda x)$$

The symbol $\Gamma(\nu)$ is called the gamma function, also referred to earlier.

The shape and scale parameters are related to the mean and variance (standard deviation squared) of the gamma distribution:

$$E[X \mid \nu, \lambda] = \mu = \nu\lambda$$

$$E\left[(X-\mu)^2 \mid \nu,\lambda\right] = \sigma^2 = \nu\lambda^2$$

Figure 5.2 shows the density for a gamma random variable with shape = 26 and scale = 3.85. Note that the mean of this distribution is 26*3.85 = 100.1.

One thing that is fairly obvious from the figure is that the gamma density is not symmetric about the mean. The lack of symmetry about the mean is referred to as skewness. Normal distributions have 0 skewness. In order to compute the probability that the gamma variable X would have values between say, a and b, we must use the CDF, namely:

$$\Pr\{a \le X \le b \mid \nu, \lambda\} = F(b) - F(a) = \int_a^b \frac{\lambda^{\nu}}{\Gamma(\nu)} x^{\nu-1} \exp(-\lambda x)\,dx$$

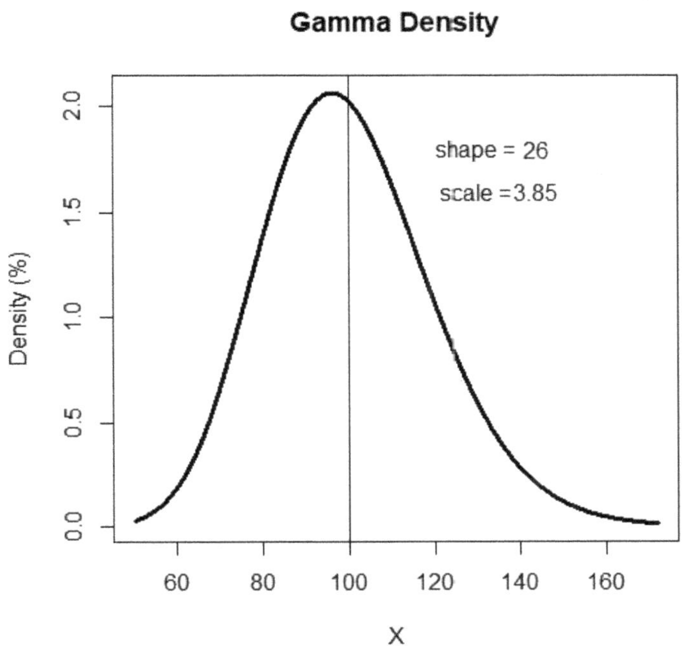

Fig. 5.2 A gamma density curve

So, for example, with $a = 70$ and $b = 130$:

$$\Pr\{a \le X \le b \mid, \nu = 26 \mid, \lambda = 3.85\} = F(130) - F(70) \approx 87.84\%$$

The gamma distribution has some special cases. If the shape parameter = 1, it is called an exponential distribution, with rate parameter $\frac{1}{\lambda}$. If its scale parameter is 1/2, it is called a Chi-squared (χ^2) distribution with ν degrees of freedom.

5.4 Estimation

The sample estimates for μ and σ from a set of values sampled from a normal population were discussed previously. That is, from the sample data, the maximum likelihood estimators for μ and σ^2 are:

$$\hat{\mu} = \frac{1}{n}\sum_{i=1}^{n} x_i$$

$$\hat{\sigma}^2 = \frac{1}{n}\sum_{i=1}^{n}\left(x_i - \hat{\mu}\right)^2$$

The maximum likelihood estimators for the gamma shape and scale parameters are a little more complicated, but still they can be computed from data:

$$\hat{\nu} = \frac{\hat{\mu}}{\hat{\lambda}}$$

and

$$\hat{\lambda} = \widehat{x\ln(x)} - \hat{\mu}\widehat{\ln(x)}$$

where:

$$\widehat{x\ln(x)} = \frac{1}{n}\sum_{i=1}^{n} x_i \ln(x_i)$$

$$\widehat{\ln(x)} = \frac{1}{n}\sum_{i=1}^{n} \ln(x_i)$$

Like the estimate for the Binomial parameter, p, these estimates, called point estimates, have some "error" associated with them. That is, different samples of size n would most likely yield different results. In order to assess the variability in point estimates, we can construct confidence intervals.

5.5 Points to Ponder

1. A sample of $n = 20$ units was drawn from a gamma distribution, and the sample mean was computed. Is it appropriate to assume that the sample mean comes from a normal distribution? Would it be appropriate to test hypotheses about the mean using critical values from a Student's t distribution? Why or why not?
2. If a gamma distribution has a shape parameter of $\nu = 1$, what is this type of distribution called?

References

Johnson, N. L., Kotz, S., & Balakrishnan, N. (1995). *Continuous univariate distributions, Vol. 1* (2nd ed.). John Wiley and Sons.

Meyer, P. L. (1970). *Introduction to probability and statistical applications* (2nd ed.). Addison-Wesley.

Chapter 6
Continuous Variables and Acceptance Sampling Plans

6.1 Sample Z-Scores and Cpk

For acceptance sampling with Binomial variables, each observation was either a "success" or a "failure," and the variable was X = the number of "successes" out of the sample size, n. But for continuous measurements/variables, there may be degrees of success or failure. There may be a range of values for the variable that are all "acceptable." The range may be bounded with a lower and upper "specification" limit, say L and U. That is, as long as the value of X falls within the lower and upper bound, the item or experimental unit on which the measurement was made is considered a "success." So why not simply convert the value of X to a binary case, namely, if X falls within the range (L, U), then count the item as a "success" (1), and otherwise, a "failure" (0)? The answer is power and sample size. To get the same power with the derived binary variable, the sample size would be close to double. So, converting the continuous variable into a binary variable is snatching defeat from the jaws of victory, so to speak.

Recall the Standard Normal variable:

$$Z = \frac{X - \mu}{\sigma}$$

We will depart from our previous notation for sample means and standard deviations. Let:

$$\bar{x} = \frac{1}{n}\sum_{i=1}^{n} x_i$$

Supplementary Information The online version contains supplementary material available at https://doi.org/10.1007/978-3-031-84479-9_6. The videos can be accessed individually by clicking the DOI link in the accompanying figure caption or by scanning this link with the SN More Media App.

S. A. Pardo, *Beyond the T-Test*, https://doi.org/10.1007/978-3-031-84479-9_6

be the sample arithmetic mean of the values $x_1, x_2, \ldots, x_n$. Then let:

$$S^2 = \frac{1}{n-1}\sum_{i=1}^{n}(x_i - \overline{x})^2$$

be the sample variance. Then:

$$\widehat{Z}_U = \frac{U - \overline{x}}{S}$$

is the sample z-score for the upper specification limit on X. Similarly:

$$\widehat{Z}_L = \frac{L - \overline{x}}{S}$$

is the sample z-score for the lower specification limit, L. In general:

$$z = \frac{x - \overline{x}}{S}$$

is the sample z-score for any arbitrary value of X.

If the sample size was extremely large, we would expect about 99.73% of the values of z to fall within the range (−3, +3). If in fact 99.73% of the values of X fell within the specification range (L, U), then 99.73% of the sample z-scores (again assuming an extremely large sample size) would fall within (−3, +3). Thus, if $\widehat{Z}_L \geq -3$ and $\widehat{Z}_U \leq +3$, then we could infer that 99.73% of the values of X would fall in the range (L, U).

Of course we are not likely to have an "extremely large" sample size. We must find some way to account for the sampling "error" in these statistics. In the case of the n Bernoulli variables, we found that their sum had a Binomial distribution. Here, instead of assuming that the observations have a Bernoulli distribution, we will assume that they are samples from a Normally distributed population with unknown parameters μ and σ. Furthermore, we are going to scale our sample z-scores, and "flip" the lower limit z-score:

$$\widehat{C}pL = \frac{\overline{x} - L}{3S}$$

and

$$\widehat{C}pU = \frac{U - \overline{x}}{3S}$$

The smaller of these two quantities is called $\widehat{C}pk$ (Bissel, 1990; Kushler & Hurley, 1992):

$$\widehat{C}pk = \min\left\{\widehat{C}pL, \widehat{C}pU\right\}$$

The population values of these statistics would be:

$$CpL = \frac{\mu - L}{3\sigma}$$

$$CpU = \frac{U - \mu}{3\sigma}$$

$$Cpk = \min\{CpL, CpU\}$$

The notion is that if the values of *X* fell within (*L*, *U*) 99.73% of the time, then (Z_L, Z_U) would fall close to within the interval (−3, +3), and $\widehat{C}pL$ and $\widehat{C}pU$ would be "close" to 1. Using the smaller of $\widehat{C}pL$ and $\widehat{C}pU$ is a "conservative" measure. That is, we are suppose the hypotheses for *Cpk* could be stated as:

$$H_0 : Cpk < 1$$

and its alternative:

$$H_1 : Cpk \geq 1$$

So the question becomes, "What is the critical value for the sample statistic $\widehat{C}pk$?" The answer lies in the "sampling distribution" of this statistic. Under the assumption that the data are a random sample from a Normal population, we have:

$$3\sqrt{n}\widehat{C}pk \sim T'\left(n-1, 3\sqrt{n}c_0\right)$$

T' represents something called a non-central *t* distribution (Johnson et al., 1995) The non-centrality parameter:

$$3\sqrt{n}c_0$$

depends on the hypothetical value of *Cpk* (in this case $c_0 = 1$). So all you need to do is to compute the 100(1-α) percentile of the non-central t distribution with $n - 1$ degrees of freedom and non-centrality $3\sqrt{n}c_0$, then divide by $3\sqrt{n}$ and you will have the critical value for $\widehat{C}pk$. As an example, if $n = 30$, $\alpha = 0.05$, and $c_0 = 1$, the non-centrality is:

$$3\sqrt{n}c_0 = 3\sqrt{30}(1) \approx 16.4317$$

The critical value can be obtained using the R qt() function:

$$\text{qt}\left(\text{p} = \text{alpha,df} = \text{n} - 1\text{,ncp} = 3^{*}\text{sqrt}(30)\right) / \left(3^{*}\text{sqrt}(30)\right) \approx 0.8041$$

Note that we are using the lower tail (5%) of this non-central *t* distribution. Thus, if the sample.

$\widehat{C}pk$ is 0.8041 or greater, we would reject the null hypothesis, H_0, in favor of the alternative H_1.

This is equivalent to saying that we would reject the hypothesis that the probability of a "success" (i.e., a value of *X* within (*L*, *U*)) is less than 99.73%. Suppose we had in fact snatched defeat from the jaws of victory and converted every observation of *X* into a binary result. That is, if *X* was within (*L*, *U*), we would call it a "success," and a "failure" otherwise. A $\widehat{Cpk}$ of 0.8041 corresponds to a sample z-score of 3(0.8041) ≈ 2.4123, which in turn corresponds to a probability of approximately 99.21%. If we had used $n = 30$ with binary variables, in order to have a critical value of more than 99%, we would have required at least 30 "success" results to PASS. Suppose the actual probability of *X* falling within (*L*, *U*) was only 95%. With the *Cpk* paradigm, there would be approximately a 1.78% chance of PASSing (i.e., getting a $\widehat{Cpk}$ of at least 0.8041). With a Binomial test, there would be approximately a 21.46% chance of PASSing (29 "successes" out of 30). In fact, to obtain the same power, the sample size for the Binomial test would need to be about 117, with a critical value of 116. That would yield about a 1.77% chance of PASSing if the actual population probability that *X* would fall within (*L*, *U*) is 95%. With the *Cpk* sampling plan an $n = 30$ gives about the same risk as the Binomial with $n = 117$.

If we formulate the hypotheses as:

$$H_0 : \Pr\{L \le X \le U\} < 0.9973$$

versus:

$$H_1 : \Pr\{L \le X \le U\} \ge 0.9973$$

then the power curves for the *Cpk* and Binomial plans are shown in Fig. 6.1.

The *Cpk* plan is more stringent (harder to PASS, easier to FAIL) than the Binomial plan, even though the sample size is much lower. That is, to have a 95% chance of PASSing for the *Cpk* plan, the $\Pr\{L \le X \le U\}$ must be approximately 99.87% and with the Binomial plan, the $\Pr\{L \le X \le U\}$ must be approximately 99.70%. Conversely, the *Cpk* plan has about a 5% chance of PASSing if $\Pr\{L \le X \le U\}$ is as low as 96.42%. The Binomial plan has about a 5% chance of PASSing if $\Pr\{L \le X \le U\}$ is as low as 96.01%. It took 3.9 times the sample size of the *Cpk* plan to get a Binomial plan to be nearly (but not quite) as powerful.

6.2 Target Values and *Cpm*

One issue with *Cpk* is that its probability content is based entirely on the values of *X* falling within the specification limits, somewhat regardless of how the population distribution is centered. However, it should be noted that if (*L*, *U*) is the range of acceptable values, perhaps there is an "ideal" value, which frequently is:

$$\tau = \frac{L + U}{2}$$

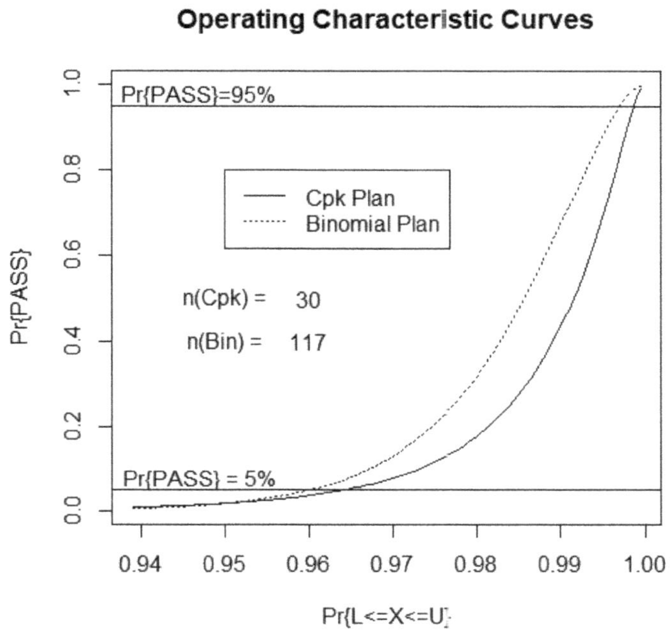

Fig. 6.1 Power curves for *Cpk* and binomial plans

It may be that there should be a means for incorporating the average departure from τ. Another index is called *Cpm* (Kushler & Hurley, 1992):

$$Cpm = \frac{U - L}{6\sqrt{(\mu - \tau)^2 + \sigma^2}}$$

If $\mu = \tau$ and $\sigma = \frac{U - L}{6}$, then $Cpm = 1$.

The associated sample statistic is:

$$\widehat{C}pm = \frac{U - L}{6\sqrt{(\bar{x} - \tau)^2 + S^2}}$$

This index incorporates the specification range (L, U) and the "target" or ideal value of X, τ.

The difference $\mu - \tau$ may be considered the "bias" between the measurement and the mean of X. The sample statistic for bias is $\bar{x} - \tau$. One nice feature of this measure is that "sample bias," $\bar{x} - \tau$, will tend toward being Normally distributed, centered around $\mu - \tau$.

Like *Cpk*, power curves for *Cpm* can also be derived or at least approximated (Boyles, 1991).

Suppose we hypothesize that:

$$\Pr\{X \le L\} = p_L$$

and

$$\Pr\{X \ge U\} = p_U$$

So that the overall probability that a value if X will be outside the limit range (L, U) would be.

$p_L + p_U$. Thus, assuming that X is normally distributed, we can compute z-scores for the specification limits:

$$Z_L = \frac{L-\mu}{\sigma} = \Phi^{-1}(p_L)$$

$$Z_U = \frac{U-\mu}{\sigma} = \Phi^{-1}(1-p_U)$$

Given that the target for X, namely τ, is known, we could find values of μ and σ that would yield the desired probabilities.

The solution for the desired values of μ and σ, call them μ^* and σ^*, are:

$$\mu^* = U\left(1 + \frac{Z_U}{Z_L - 1}\right) - L\left(\frac{Z_U}{Z_L - 1}\right)$$

$$\sigma^* = -U\left(\frac{1}{Z_L - 1}\right) + L\left(\frac{1}{Z_L - 1}\right)$$

Thus, the z-score for the target is:

$$Z_\tau = \frac{\tau - \mu^*}{\sigma^*}$$

Ideally, $\tau = \mu^*$. If p_L does not equal p_U, then μ^* may not equal μ.

Let $\lambda = nZ_\tau^2$ and $\nu = n$. Then Boyles has found that $\widehat{C}pm$ is approximately distributed as:

$$\widehat{C}pm \sim \sqrt{\frac{\nu}{\chi^2}}C_0$$

where C_0 is the hypothesized value of Cpm and χ^2 is a non-central Chi-squared variable with ν degrees of freedom and non-centrality parameter:

$$\lambda = nZ_\tau^2$$

Let $P = 1 - (p_L + p_U)$ = the "null" probability tha: the measurement is within the interval $[L, U]$.

The value of C_0 is, per hypothesis (Perakis & Xe:kalaki, 2016) is:

$$C_0 = \frac{1}{3}\Phi^{-1}\left(\frac{1+P}{2}\right)$$

The hypotheses can be stated as:

$$H_0 : Cpm < C_0$$

versus:

$$H_1 : Cpm \geq C_0$$

The critical value for $\widehat{C}pm$ is:

$$\widehat{C}pm_{crit} \sim \sqrt{\frac{\nu}{\chi^2_{n,1-\alpha,nZ_\tau^2}}}C_0$$

and $\chi^2_{n,1-\alpha}$ is the 100(1 − α) percentile of a non-centr:l Chi-Squared distribution with n degrees of freedom and non-centrality $\lambda = nZ_\tau^2$. When $\tau = \mu$, $Z_\tau = 0$, so the non-centrality is also 0, making this a "central" Chi-Squ:red.

If the sample *Cpm* exceeds $\widehat{C}pm_{crit}$, then the null hypothesis, H_0, is rejected in favor of the alternative, H_1.

Consider an example:

$n = 30$
$\mu^* = 100$
$\sigma^* \approx 0.3333$
$C_0 = \frac{1}{3}\Phi\left(\frac{1+.9973}{2}\right) \approx 1.00$
$L = 99$
$U = 101$
$\tau = 100$
$p_L = 0.00135$
$p_U = 0.00135$
$\alpha = 0.05$

The critical value for $\widehat{C}pm$ is approximately 0.8279. If $\widehat{C}pm$ is at least 0.8279, the plan PASSes. Fig. 6.2 ` Notice how as the target moves away from μ, its z-score gets larger in absolute value and the probability of PASSing decreases.

Boyles (1991) provided a formula for a lower confidence bound on *Cpm*:

$$Cpm(1-\alpha) = Cpm\sqrt{\frac{\chi^2_{c,\upsilon}}{\upsilon}}$$

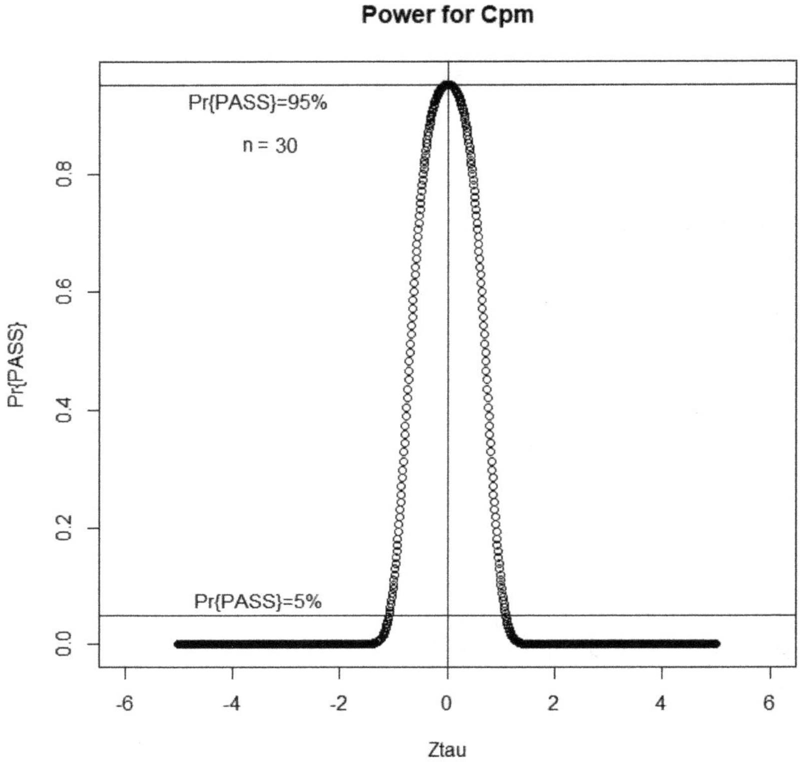

Fig. 6.2 Power curve for *Cpm*

with

$$\hat{\lambda} = n\left(\frac{\overline{x} - T}{s}\right)^2$$

and

$$\upsilon = \frac{\left(n + \hat{\lambda}\right)^2}{n + 2\hat{\lambda}}$$

That is, $\hat{\lambda}$ is an estimate of the non-centrality parameter and υ is the degrees of freedom.

6.3 Root Mean Square Error (RMSE)

A statistic related to *Cpm* is what some call root mean square error (RMSE). If, as in *Cpm*, τ represents the target value for variable *X*, then:

$$\text{MSE} = \frac{1}{n}\sum_{i=1}^{n}(x_i - \tau)^2$$

is the mean square error. Its square root is the RMSE. If:

$$\text{SE} = \sum_{i=1}^{n}(x_i - \tau)^2 = \sum_{i=1}^{n}(x_i - \mu + \mu - \tau)^2$$

and if X has a normal distribution with mean μ and standard deviation σ, then:

$$\sum_{i=1}^{n}(x_i - \tau)^2 = \sum_{i=1}^{n}(x_i - \mu + \mu - \tau)^2 \sigma^2 \sum_{i=1}^{n}\left(\frac{x_i - \mu}{\sigma} + \frac{\mu - \tau}{\sigma}\right)^2$$
$$\sim \sigma^2 \chi'^2\left(n, \lambda = \frac{n(\mu - \tau)^2}{\sigma^2}\right)$$

That is, SE has a non-central Chi-squared distribution and MSE is distributed as:

$$MSE = \frac{1}{n}\sum_{i=1}^{n}(x_i - \tau)^2 \sim \frac{\sigma^2}{n}\chi'^2\left(n, \lambda = nZ_\tau^2\right)$$

$$Z_\tau = \frac{\tau - \mu}{\sigma}$$

where λ is the non-centrality parameter. Thus, RMSE is distributed as a non-central Chi:

$$\text{RMSE} = \sqrt{\frac{1}{n}\sum_{i=1}^{n}(x_i - \tau)^2} \sim \frac{\sigma}{\sqrt{n}}\chi'\left(n, \lambda' = \sqrt{n}\left|Z_\tau\right|\right)$$

The RMSE has the same units as X. So keep that in mind when formulating a null hypothesis. Also keep in mind that while large values of *Cpm* are desirable, smaller values of RMSE are better than larger values.

If the null hypothesis is stated as:

$$H_0 : \text{RMSE} > rmse_0$$

the alternative (which is the desired state) is:

$$H_1 : \text{RMSE} \le rmse_0$$

The null value of RMSE is its expected value:

$$rmse_0 = \sqrt{\left(\mu^* - \tau\right)^2 + \sigma^{*2}}$$

Assuming that there are two specification limits, L and U, then μ^* and σ^* are defined the same way that they were for *Cpm*:

$$\mu^* = U\left(1 + \frac{Z_U}{Z_L - 1}\right) - L\left(\frac{Z_U}{Z_L - 1}\right)$$

$$\sigma^* = -U\left(\frac{1}{Z_L - 1}\right) + L\left(\frac{1}{Z_L - 1}\right)$$

Of course, this also assumes that there are "null" values for p_L, and p_U, the probabilities of a measurement falling outside the specification limits.

The critical value for RMSE would be chosen such that:

$$\Pr\left\{\widehat{RMSE} > rmse_c | RMSE = rmse_0\right\} = \alpha$$

where α is a small probability, most often chosen to be 0.05.

With $n = 30$, $\tau = 100$, $L = 99$, $U = 101$, $p_L = p_U = 0.00135$, $\mu^* = 100$, $\sigma^* = 0.3333$, the critical value is $rmse_c \approx 0.4026$. The power curve is shown in Fig. 6.3.

RMSE is a measure of deviation from target. *Cpm* is also a measure of deviation from target. They both incorporate specification limits as well at the target value, whereas *Cpk* only is concerned with having values fall with limits, whether is it is

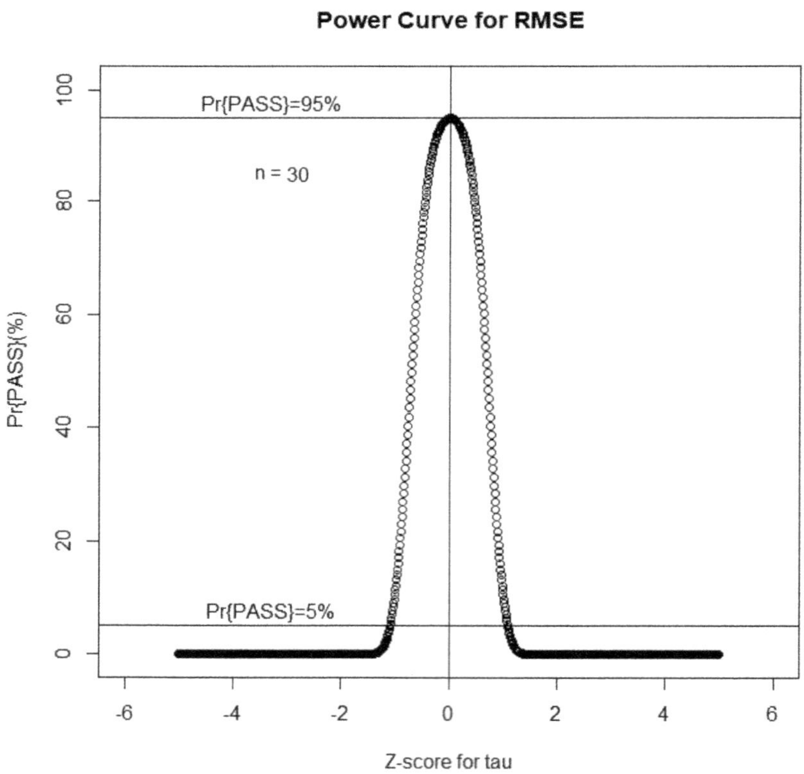

Fig. 6.3 Power curve for RMSE

due to the population mean falling well within the limits or the population standard deviation is small enough to compensate for "bias" between the target and the mean. They all are powerful methods for acceptance sampling.

6.4 Summary

Like any hypothesis-testing situation, acceptance sampling is a game of belief. We hypothesize something about a parameter, and then use data to decide whether we should either believe the "null" hypothesis or its logical negation. However, there is always the possibility that we might make the wrong decision. The key to successful acceptance sampling is the power curve. It tells you how likely it is to make the incorrect decision under various possibilities for the truth. Even when the truth matches the hypothesis, there is some chance that the wrong decision will be made. Usually that chance is fixed to 5%, but even then, it is only as good as the assumptions, such as the normality of the population or the probability of a success not changing as samples are obtained. Still, we are faced with the problem that we must make a decision, so we would at least like to know the range of risk associated with making the wrong one. Keep in mind, however, that regardless of what results are obtained, the actual truth about the population is generally never known.

6.5 Points to Ponder

1. Suppose a quality engineer told you that a sampling plan with $n = 50$ had a power of 95%. Is this sufficient information to justify the sample size? If not, what other information would you want?
2. Do you see any advantage of either *Cpm* or RMSE when a target value for a random variable relating to product quality is of interest? Why or why not?
3. When is it appropriate to use *Cpk* as the measure of goodness?

References

Bissel, A. F. (1990). How reliable is your capability index? *Applied Statistics, 39*(3), 331–340.

Boyles, R. A. (1991). The Taguchi capability index. *Journal of Quality Technology, 23*(1), 17–26.

Johnson, N. L., Kotz, S., & Balakrishnan, N. (1995). *Continuous univariate distributions, Vol. 2* (2nd ed.). John Wiley and Sons.

Kushler, R. H., & Hurley, P. (1992). Confidence bounds for capability indices. *Journal of Quality Technology, 24*(4), 188–195.

Perakis, M., & Xekalaki, E. (2016). On the relationship between capability indices and the proportion of conformance. *Journal of Quality Technology and Quantitative Management, 13*(2), 207–220.

Chapter 7
Confidence

7.1 What Is Confidence

Confidence is an English word meaning, among other things: TRUST, RELIANCE, or SELF-ASSURANCE, BOLDNESS (Merriam-Webster, 2005). Well, the word "confidence" is used in the science of statistics to mean something very technical. It means:

> The probability that an interval* (a range of numerical values), constructed from empirical observations, contains the actual, but unknown, value of some particular parameter or set of parameters governing the population's distribution of values from which those empirical observations are sampled.

This is the definition (more or less) formulated by Jerzy Neyman, back in the 1920s (Armitage, 1971). Other definitions contended for such (and related) intervals, most notably the "fiducial limits" of Sir Ronald A. Fisher, Sc.D., FRS. However, Professor Neyman's definition won the prize.

Wow. So much for Merriam-Webster. This statistical definition requires some unpacking.

First, what we mean by "values" must be understood. "Values" refer to (quantified) observations or measurements of random variables. Some examples of random variables are the "fasting" (say 9 h after eating last) blood glucose (FBG) of adults with Type 1 diabetes, the shoe size of children, the color of a person's hair (say, 1 = blonde, 2 = brown, 3 = red), the fuel consumption rate of the Abrams M-1 Main Battle Tank, or the maximum speed of the Boeing 767 aircraft. For the last two variables, assume that any two vehicles of the same type may have different results, due to variation in the manufacturing process.

Supplementary Information The online version contains supplementary material available at https://doi.org/10.1007/978-3-031-84479-9_7. The videos can be accessed individually by clicking the DOI link in the accompanying figure caption or by scanning this link with the SN More Media App.

S. A. Pardo, *Beyond the T-Test*, https://doi.org/10.1007/978-3-031-84479-9_7

The next part is the notion of "population." As first discussed, a population is the entire collection of all individuals upon which we could make the measurement of some particular random variable. So, for the fasting blood glucose example, our population might be all adults who have Type 1 diabetes, whether they are alive now, were alive at one time, or have yet to be born and grow into adults. You might say, "But, you cannot measure the fasting blood glucose of people who are no longer living or of those people yet to be…" While this is true, the point is that those people either did or will have a fasting blood glucose, assuming that there was at least one 9 h period in which they consumed (or will consume) no food.

For the children's shoe size example, similarly we could talk about all children who ever existed or who may exist in the future. The point is that a population is a complete collection (possibly of infinite number) of individuals having some things in common, where those things are the factors that we use to define the population.

Random variables will have values differing from one individual to the next within the population of interest. The frequency distribution, or just distribution, is the probability that the random variable can have values falling within various intervals. The range of values for a random variable could be infinite, half-infinite, or completely bounded. So, for example, fasting blood glucose cannot be a negative number, but there is no defined upper boundary. Thus, the distribution of fasting blood glucose is half-infinite, since fasting blood glucose can never be less than zero. Now, you might argue that the distribution of fasting blood glucose is really bounded, because no living human (or human who was once living, or human who will come into existence and grow to adulthood) could have a fasting blood glucose value of, say, 10,000 mg/dL (555.06 mmol/L). While that may be true, since no one could define the exact upper bound, above which it is impossible to find any value, we will treat fasting blood glucose as half-infinite. What we can do is to say that the interval of values from, say 2000 mg/dL to positive infinity ($+\infty$) has a very, very low probability.

While we are making definitions, let us loosely define probability to mean the "frequency of occurrence."

So populations can have random variables defined as some measurement or quantified observations having values for each individual in the population, and these values in turn have a distribution.

The next part of the confidence definition we will discuss is the word "parameter.". As seen in the cases of the Binomial, Normal, and Gamma random variables, a distribution can be described by a curve, or function, that shows something about frequency of occurrence over the entire range of possible values for the random variable of interest. This curve can often be described by a mathematical function that has some set of fixed parameters or numbers that describe the shape of the curve. For example, the probability of a "success" for the Binomial, the mean and variance of the Normal, or the shape and scale parameters for the Gamma. In fact, even if we do not know the form of this mathematical function, we can postulate that it has certain parameters, that would govern its shape. We may be more interested in the values of those parameters, and not so much interested in the general shape of that distribution function. One such parameter we are commonly interested in is the mean, or average value.

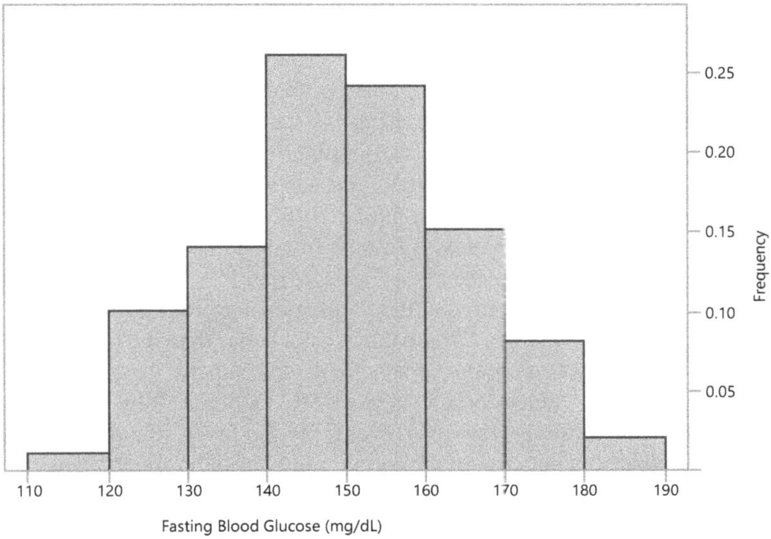

Fig. 7.1 Histogram of n = 100 fasting blood glucose (FBG) measurements

The statistical problem is that we do not know the values of those parameters, but we wish we did. So, for example, consider the frequency curve for average fasting blood glucose. Suppose we sampled n = 100 adults with Type 1 diabetes (never mind about how the sample is drawn, for the time being), and for each of those individuals, we measured a fasting blood glucose (once per individual). We could make a histogram of the values we observed. Figure 7.1 shows such a histogram.

We see a frequency plot, where we computed the proportion (frequency) of values observed between 110 and 120 mg/dL, 120–130 mg/dL, …, and 180–190 mg/dL. This histogram was generated using measurements from only n = 100 individuals sampled from a population of effectively an infinite number of individuals. Suppose we wanted to know the average FBG of this population. The average value is a parameter of the population's FBG distribution. Well, since we have only n = 100 values, we cannot actually know the average FBG for the whole population. Rather, all we can do is estimate that parameter. We have seen that the "best" way to estimate the population mean (maximum likelihood) is the sample's arithmetic average, which in this case is 150.4 mg/dL. So, is 150.4 mg/dL the average FBG of ALL adults with Type 1 diabetes? If we got a different sample of people, or even if we used the same 100 people, but had them repeat the experiment a month later, would we get the same sample average?

What we would like to do is find a "feasible" range of values in which we are fairly certain (confident) that the population's mean FBG falls into. With a sample size of n = 0 (!!!!) we can say with absolute certainty (100% confidence) that the population's average FBG falls somewhere between 0 and +∞ mg/dL. That is 100% confidence! How much better can you get? Well, you might say, "excuse me, but could you narrow down that range just a bit?". Well, we can, but there are two prices to pay:

1. We will have to sample more than $n = 0$;
2. We will have to give up on having 100% confidence.

Instead of 100% confidence, we will only be able to say we are "fairly certain." We will define quantitatively what we mean by "fairly certain" in a little bit. The narrowness of our "feasible range" for the population parameter depends mostly on the sample size, n. The bigger the sample, the narrower the range. Recall that the standard error of the sample mean gets smaller as the sample size gets larger. In this case, we can use some mathematical formulae (not specified here) to find that our feasible range for being fairly certain is 147.5 mg/dL to 153.3 mg/dL. That is, based on these data, we are fairly certain that the population's average FBG is somewhere between 147.5 and 153.3 mg/dL. OK, so that is a lot better than the interval $(0, +\infty)$. But, we gave up the 100% confidence statement. Instead, we can only say "fairly certain." Quantitatively, "fairly certain" means 95% confident. Why 95%? Well, it is ultimately somewhat arbitrary. However, ever since the 1890s (thanks to Sir Gilbert Walker, who studied the patterns of Monsoon winds in India for the British Royal Meteorological Service), experimenters and data analysts have conventionally accepted a 5% (1 out of 20) risk of being wrong about the interval in which the population parameter actually lies. If we wanted, say, 99% confidence, our interval would be wider (146.5, 154.3 mg/dL). If we wanted only 90% confidence, our interval would be narrower (147.9, 152.8 mg/dL). So, choosing 95% confidence is a trade-off between specificity (narrowness: (147.5, 153.3 mg/dL) is narrower than (146.6, 154.3 mg/dL)) of the interval, and risk of missing the true value (risk: 100%–95% = 5% is less risk than 100%–90% = 10%). Sometimes we are only interested in one side of a confidence interval. For example, we might only want to know the lower confidence bound on the probability of "success"; in that case, the upper bound is not interesting, because we would be perfectly happy if the probability of success was 100%, but we may want to know the worst it could "feasibly" be, based on the data we have gathered. Regardless of whether you want a two-sided interval or a single bound or limit, use any confidence level you want as long as it is 95%. OK, there are some special circumstances under which 90% or 99% confidence levels should be used, but we will not discuss those circumstances in this document.

We must understand that confidence intervals are constructed from empirical observations, or data, from a finite sample of individuals. The sample is a subset of the population. If we took a different sample, we would most likely get a slightly different interval. The point is that generally we cannot observe every single member of a population, nor can we continually sample and construct many intervals. We must get as much information as we can from our single sample. The confidence interval gives us such information, and quantifies it with a confidence level.

Earlier we alluded to the way in which a sample of individuals is selected. The simplest paradigm is called random sampling. That means every individual in the population of interest is equally likely to be selected to be part of the sample. Sampling paradigms can be very complex, and affect the formulas used to compute confidence intervals. We, however, are not going to discuss any of that here. In fact, at this point, we are only assuming that the sample was obtained in some legitimate fashion, so that it is possible to compute a confidence interval without invalidating any necessary assumptions.

7.2 What Confidence Does NOT Mean

The term "confidence" can ONLY be applied to an interval (whether it is bounded or unbounded on one or both sides). It makes sense, for example, to say: "We are 95% confident that the probability a blood glucose (BG) meter has ±15% error is between 98.5% and 99.5%." It would also make sense to say, "We are 95% confident that there is at least a 95% probability of meter errors falling within ±15% of lab results." Note that the phrase "at least" implies an interval, namely (95%, 100%).

It does not make sense to say, "We are 95% confident that the average meter error is 8.5%." In this statement, the value 8.5% is not an interval. It also does not make sense to ask the question, "What sample size will give us 95% confidence to pass the test?" You can ask "What sample size will make the width of a confidence interval for the probability of success to be no more than ±1%?" The width of the confidence interval is the difference between the lower bound to the upper bound of the interval. More commonly we refer to the half-width (e.g., ±1%, and not the full width of 2%). The width of the confidence interval is mostly dependent on the sample size. The confidence level (e.g., 90%, 95%, 99%) also affects the width, but it also affects the interpretation (i.e., the risk of missing the value of the population parameter).

Likewise, it does not make sense to ask, "What confidence do we have of passing the test if there is a 90% probability of success on each attempt?" Confidence is a probability that is applied after data have been gathered, and confidence can only apply to an interval. We could make confidence statements about a hypothetical situation. For example, we could say, "If we sample $n = 100$ people, and find that 70 of them are right-handed, then we would be 95% confident that the percent of right-handed people in the population is somewhere between 60% and 79%.". You can ask the statistician to find a sample size that would yield a particular interval width. Or you can ask the statistician to tell you, for a given sample size, the smallest value of a sample statistic (e.g., the sample arithmetic average, or the percent of individuals satisfying a particular criterion) required in order for the lower/upper confidence limit (lower/upper bound on a confidence interval) to have some particular value.

The probability of passing or failing a statistical test of some given hypothetical condition or value of an unknown parameter is called "power." It is often mistaken for confidence. Don't make that mistake. Power is the probability of concluding from data that some hypothesis about a parameter is incorrect. Confidence is the degree of feasibility that an interval has for containing the actual value of the parameter. We can determine a sample size before performing an experiment (a priori) that would give us a particular power to reject a hypothesis given it is "false," but we cannot compute a sample size to give us some particular level of confidence about a parameter's value. We can, however, determine a sample size that would yield a confidence interval with some specific confidence level, and a specified width. It is true that there is a relationship between power and confidence, but they are not the same, and their relationship is a topic for another day.

In summary, confidence intervals are feasible ranges for the actual value of a population parameter. Confidence intervals have an associated confidence level, which is a probability that indicates the degree to which we can be certain the interval actually contains the true (but unknown) value of the parameter. The confidence level ought to be 95%, except in some special circumstances. Confidence intervals are constructed after data are collected (a posteriori). The sample size governs the specificity (width) of the interval (although confidence level also has an effect).

We speak about intervals in one dimension, but confidence regions can be constructed in a multidimensional parameter space situation. However, for now we are restricting ourselves to an interval for a single "scalar" parameter, and not a parameter vector.

7.3 Some Consequences of Confidence Limits

When we compute some value, such as a mean, standard deviation, or proportion, from sample data, the results of those computations are known as "point" estimates of a corresponding population parameter. So, for example suppose that is a sample of n = 100 results, 95 of them are considered to be "good" or "successful," and 5 of them not so much. Our point estimate for the probability of obtaining a "good" result on a new "try" is:

$$0.95 = \frac{95}{100} = 95\%$$

We know that this value, 0.95, may not remain if we samples yet another 100 "tries." The confidence interval will be a range of values, starting with something below 0.95 (lower limit) and bounded above by some value greater than 0.95. In the example of $n = 100$, with 95 "good" results, the 95% confidence interval for the probability of observing a "good" result is about (0.8998, 0.9834). Clearly, although the point estimate for this probability is 0.95, the lower confidence limit is 0.8998, which is less than 0.95. Similarly, the upper limit is 0.9834, which is greater than 0.95. Thus, if one desires the lower limit of a confidence interval to be above some value, then the point estimate must be greater than that desired lower bound. In this case, if we desired the lower confidence limit to be greater than 0.95, then the smallest point estimate we would need with $n = 100$ is 0.99. That is, to have the lower confidence limit for a 95% confidence interval to be greater than 0.95 with $n = 100$, we would have to observe at least 99 "good" results. The larger the sample size, the smaller the range of the confidence interval. So, if instead of $n = 100$, we had $n = 1000$, then we would need to observe at least 963 (96.30%) "good" results in order to have the lower confidence limit to be greater than 0.95. The 95% confidence interval would be (0.9505, 0.9734). Clearly, 0.963 falls within this interval.

Throughout the above discussion, we referred to confidence intervals have a finite lower and upper limit. It is possible to have confidence intervals where one of the limits is either +∞ or -∞. Such intervals are referred to as "one-sided." In the case of percentages (or proportions), the "infinite" limit would be truncated at either 0 or at 1. So, for example, suppose we observed n = 100 "tries," which resulted in 95 "good" values. The lower one-sided 95% confidence interval for the probability of obtaining another "good" result would be (0.9099, 1.000). That is, the lower 95% confidence limit is 0.9099, or 90.99%. We would usually refer to the result 90.99% as simply the one-sided 95% lower confidence limit.

7.4 Points to Ponder

1. What is the difference between confidence and power?
2. Are confidence statements relating to population parameters or sample statistics?

References

Armitage, P. (1971). *Statistical methods in medical research, 4th printing*. Blackwell Scientific Publications.

Merriam-Webster. (2005). *The Merriam-Webster dictionary, new edition*. Merriam-Webster.

Chapter 8
Some Confidence Interval Computations, Including Bootstrapping

8.1 Means

As noted earlier, the quantity:

$$t = \frac{\hat{\mu} - \mu}{\hat{\sigma} / \sqrt{n}}$$

has a Student's t distribution with either n or $n - 1$ degrees of freedom, depending on whether n or $n - 1$ was used to compute the estimate of standard deviation, $\hat{\sigma}$. For now, we will assume that $n - 1$ was used. But the value of μ is unknown. When testing hypotheses about μ, we can use the hypothetical value, call it μ_0, to decide in the data indicate we should believe that is a reasonable value for μ. That is, if:

$$|t| = \left| \frac{\hat{\mu} - \mu_0}{\hat{\sigma} / \sqrt{n}} \right| > t_{1-\frac{\alpha}{2},n-1}$$

Then we will NOT believe the hypothesis:

$$H_0 : \mu = \mu_0$$

$t_{1-\frac{\alpha}{2},n-1}$ is the $100\left(1 - \frac{\alpha}{2}\right)^{\text{th}}$ percentile of a Student's t distribution with $n - 1$ degrees of freedom. We use absolute values ($|t|$), since the t distribution is symmetric about 0, i.e.,

---

Supplementary Information The online version contains supplementary material available at https://doi.org/10.1007/978-3-031-84479-9_8. The videos can be accessed individually by clicking the DOI link in the accompanying figure caption or by scanning this link with the SN More Media App.

S. A. Pardo, *Beyond the T-Test*, https://doi.org/10.1007/978-3-031-84479-9_8

$$t_{1-\frac{\alpha}{2},n-1} = \left| t_{\frac{\alpha}{2},n-1} \right|$$

So what do we do if we do not have any μ_0 in mind. Solving the t equation for μ yields:

$$\mu = \hat{\mu} \pm t_{1-\frac{\alpha}{2},n-1} \frac{\hat{\sigma}}{\sqrt{n}}$$

Note that we equated the statistic with two percentiles of the t distribution:

$$\frac{\hat{\mu} - \mu}{\hat{\sigma} / \sqrt{n}} = t_{1-\frac{\alpha}{2},n-1}$$

and

$$\frac{\hat{\mu} - \mu}{\hat{\sigma} / \sqrt{n}} = t_{\frac{\alpha}{2},n-1} = -t_{1-\frac{\alpha}{2},n-1}$$

So, when you solve these two equations for μ you get two values (hence the ± symbol):

$$\mu_{L,U} = \hat{\mu} \pm t_{1-\frac{\alpha}{2},n-1} \frac{\hat{\sigma}}{\sqrt{n}}$$

These two values comprise the $100(1 - \alpha)\%$ confidence interval for μ. In the case of the confidence interval for the mean, the interval is symmetric about the estimate $\hat{\mu}$. It is possible to have an asymmetric interval. That is, you could equate the statistic to two different values from the t distribution, say α_1 and α_2, so that the total confidence level is $\alpha_1 + \alpha_2$. Commonly, however, the interval is kept symmetric.

As an example, suppose $n = 50$, $\alpha = 0.05$, $\hat{\mu} = 100$, and $\hat{\sigma} = 10$. The symmetric $100(1 - \alpha)\% = 95\%$ confidence interval is:

$$\mu = 100 \pm 2.0096 \frac{10}{\sqrt{50}} \approx (97.158, 102.842)$$

So now we can say we are 95% confidence that the population mean lies somewhere between 97.158 and 102.842. The goodness of this information is context-sensitive. Are we glad that this interval is nearly certain of capturing the population mean? Without a context, we cannot know. What we can say is that if the experimenter says, "Well, which is it? Is the mean 97.158 or 102.842? It matters which one it is." If that is the question, then we know the sample size was not adequate. If, on the other hand, the experimenter says, "I do not care if the mean is 97.158 or 102.842. Either of those possibilities have the same consequence or implication." In that case, we would know that the sample size was adequate.

8.2 Proportions

There are three different formulas presented here. Both are approximations, albeit pretty good ones.

The first is the Clopper and Pearson (1934) method. The formulas for lower and upper confidence limits are:

$$P_L = \frac{X * F_{\alpha/2}^{-1}\left(2X, 2(n - X + 1)\right)}{n - X + 1 + X * F_{\alpha/2}^{-1}\left(2X, 2(n - X + 1)\right)}$$

and:

$$P_U = \frac{(X + 1) * F_{1-\alpha/2}^{-1}\left(2X, 2(n - X)\right)}{n - X + (X + 1) * F_{1-\alpha/2}^{-1}\left(2(X + 1), 2(n - X)\right)}$$

P_L = lower limit.
P_U = upper limit.
X = number of events or results of interest (e.g., test is positive or test is negative).
$F_q^{-1}(num, denom)$ = 100qth percentile of an F distribution (i.e., inverse F) with numerator degrees of freedom *num* and denominator degrees of freedom *denom*.

The second method is from Wilson (1927):
The $(1 - 2\alpha)100\%$ confidence interval, (L.L., U.L.), is:

$$\text{L.L.} = \frac{\hat{p} + \frac{z_\alpha^2}{2n} + z_\alpha\sqrt{\frac{\hat{p}(1-\hat{p})}{n} + \frac{z_\alpha^2}{4n^2}}}{1 + \frac{z_\alpha^2}{n}}$$

$$\text{U.L.} = \frac{\hat{p} + \frac{z_{1-\alpha}^2}{2n} + z_{1-\alpha}\sqrt{\frac{\hat{p}(1-\hat{p})}{n} + \frac{z_{1-\alpha}^2}{4n^2}}}{1 + \frac{z_{1-\alpha}^2}{n}}$$

$$\hat{p} = \frac{X}{n}$$

The z values are percentiles from a standard normal distribution (i.e., z-scores).

There is a formula that is mathematically the same as the Clopper-Pearson, but it looks very different. That formula (Scholtz, 2019) is:

$$PB_L = B^{-1}\left(\frac{\alpha}{2}, shape = X, scale = n - X + 1\right)$$

$$PB_U = B^{-1}\left(1-\frac{\alpha}{2}, shape = X+1, scale = n-X\right)$$

$B^{-1}(p)$ is the inverse distribution function of a Beta distribution with parameters α = shape and β = scale. The inverse beta distribution function can be computed with the R function *qbeta(p,shape1,shape2)*. The *shape1* parameter is what we have previously called "shape" and the *shape2* parameter we called "scale."

8.3 Standard Deviations

A 100(1 − α)% upper confidence limit for standard deviation is:

$$\sigma_U = \sqrt{\frac{\chi^2_{1-\alpha,df}}{\hat{\sigma}^2}}$$

$\chi^2_{1-\alpha,df}$ is the 100(1 − α)% percentile of a (central) Chi-Squared distribution with *df* degrees of freedom. Note that *df* is either n or $n-1$, depending on how $\hat{\sigma}^2$ was calculated.

8.4 Bootstrap Confidence Limits

Many times a parameter of interest has no special formula for confidence limits. Efron (1982) describes a method for computing a confidence interval for virtually any parameter that can be estimated with data. The method we will discuss is called "bootstrapping." It is an iterative method using random sampling of data. This is the procedure:

1. First obtain a sample of n values from the population and random variable of interest.
2. Randomly sample n values from the n sample data values, with replacement. That means you may sample the same value more than once.
3. With this "resampled" group of data, compute the parameter estimate of interest. Store it.
4. Repeat #1 and #2 about 2000 times.
5. Find the $100\frac{\alpha}{2}$ and $100\left(1-\frac{\alpha}{2}\right)$ percentiles of the distribution of the 2000 estimates. Those comprise the limits of the 100(1 − α)% confidence interval for the parameter.

This procedure is referred to as the non-parametric bootstrap percentile method.

As an example of parameters that have no explicit formulae for confidence limits, consider the 75th percentile and the scale parameter of a gamma distribution.

Suppose we sampled $n = 100$ values from a gamma distribution that actually has a shape parameter of 4.5 and a scale parameter of 40. The actual 75th percentile of this distribution is approximately 227.78.

Using a bootstrap approach, we can compute 2000 estimates of the 75th percentile and 2000 estimates of the scale parameter. Recall that the maximum likelihood estimate of the scale parameter is:

$$\hat{\lambda} = \widehat{xln(x)} - \hat{\mu}\widehat{\ln(x)}$$

The carats here on the right side of this equation stand for arithmetic averages. That is:

$$\widehat{\ln(x)} = \frac{1}{n}\sum_{i=1}^{n} \ln(x_i)$$

$$\widehat{\mathrm{x}\ln(x)} = \frac{1}{n}\sum_{i=1}^{n} \mathrm{x}_i \ln(x_i)$$

and as done earlier:

$$\hat{\mu} = \frac{1}{n}\sum_{i=1}^{n} x_i$$

If we resampled 2000 times a sample of $n = 100$ values, we could compute 95% confidence intervals for the 75th percentile and the scale parameter.

Table 8.1 provides the results.

Notice that both confidence intervals do contain the true parameter values. Fig. 8.1 shows the histogram of the 2000 bootstrap estimates of the 75th percentile and Fig. 8.2 shows the histogram for the scale estimates. Recall that the median in the 50th percentile of the 2000 bootstrap estimates.

A fairly common situation in which bootstrapping can be quite useful is when the desired parameter is the difference between two proportions. That is, suppose that there are either two conditions or two groups. That is, suppose the experimenter is interested in whether the probability of a "success" is different for two different

Table 8.1 Bootstrap results for Gamma example

(Code 8.1)

Parameter	True Value	Sample Size, n	No. Bootstrap Resamplings	Bootstrap Median	LCL	UCL
75th Percentile	227.78	100	2000	237.75	209.90	264.46
Scale	40.00	100	2000	32.34	24.49	43.09

(LCL,UCL) = 95% Bootstrap Percentile Method Confidence Intervals

(Code 8.1)

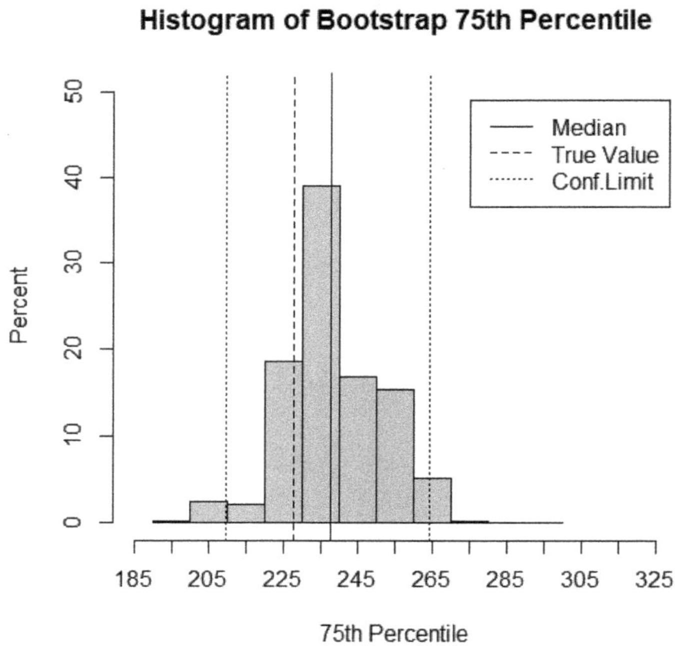

Fig. 8.1 Histogram of bootstrap 75th percentile estimates

conditions. It might be desired to construct a confidence interval for the difference in the probability of a success between the two conditions:

$$\Pr\{success|condition1\} - \Pr\{success|condition2\}$$

There are some approximate formulas for such confidence intervals, but they can be very inaccurate. Another way to compare the proportion of successes between two groups is via odds ratios. The odds of success are:

$$O = \frac{\Pr}{1 - \Pr}$$

where *Pr* is the probability of success. An odds ratio is the odds of success for one group divided by the odds of success for the other group.

Of course, in a sample of n binary results, where a "success" is coded as a 1 and a non-success as a 0, and X = the number of 1's, the sample proportion is:

$$p = \frac{X}{n}$$

The sample odds is:

(Code 8.1)

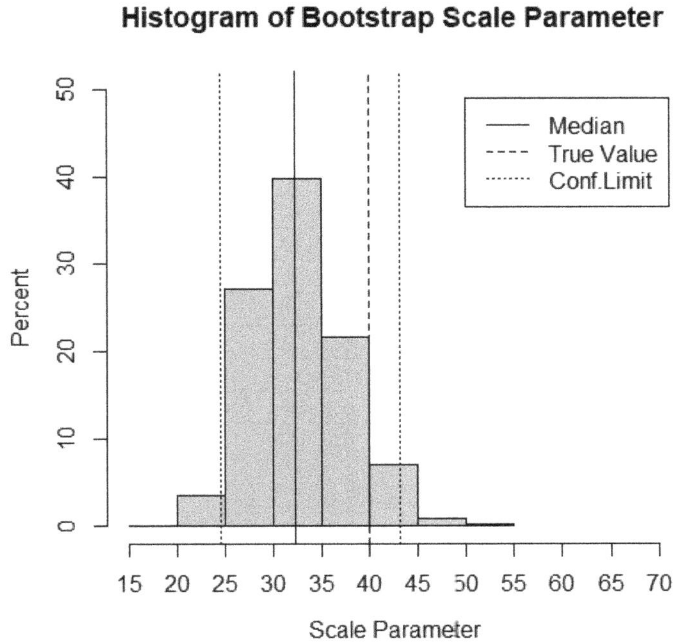

Fig. 8.2 Histogram of bootstrap scale estimates

$$o = \frac{p}{1-p}$$

So if there are two groups, A and B, then:

$$p_A - p_B$$

and

$$or = \frac{o_A}{o_B}$$

If the confidence interval for the difference contains 0, then there is no significant difference between the two groups. Similarly, if the confidence interval for odds ratio contains 1, there is no significant difference between the two groups.

There are no good formulas for confidence intervals for either of these quantities, but bootstrap approach can be every effective.

Two groups were sampled, where each unit was judged as either a "success" (1) or "non-success" (0). The proportions of successes were tallied out of $n = 100$ for both. The odds and odds ratio were all computed. Bootstrap samples were taken, and the differences and odds ratios computed for $N = 2000$ repetitions. The

Table 8.2 Results from bootstrap sampling of PA – PB and Odds.A/Odds.B

(Code 8.2)

Quantity	Value
Sample Size per Group	100
prop. "success" Group A	0.95
prop. "success" Group B	0.86
Diff: PA - PB	0.09
Odds of "success", Group A	19.0000
Odds of "success", Group B	6.1429
Odds ratio: A / B	3.0930
LCL, Diff	0.01
UCL, Diff	0.17
LCL, OR	1.1548
UCL, OR	13.6364

(LCL, UCL) = Two-Sided 95% Confidence Interval

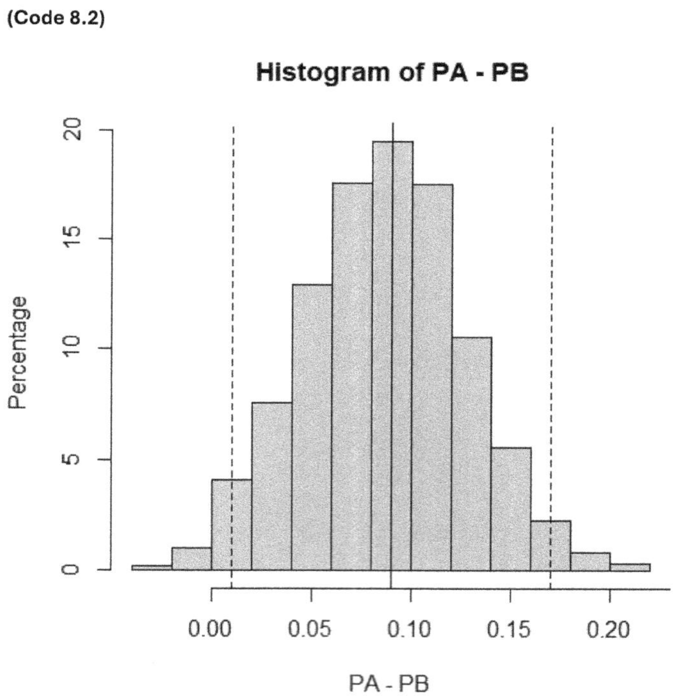

Fig. 8.3 Histogram of bootstrap differences: PA – PB

two-sided 95% confidence intervals for both the differences in the proportions of successes and their odds ratios were computed. Table 8.2 provides the results.

Figure 8.3 shows the histogram of the bootstrap samples of the difference in proportions, and Fig. 8.4 shows the histogram of bootstrap samples of OR. The

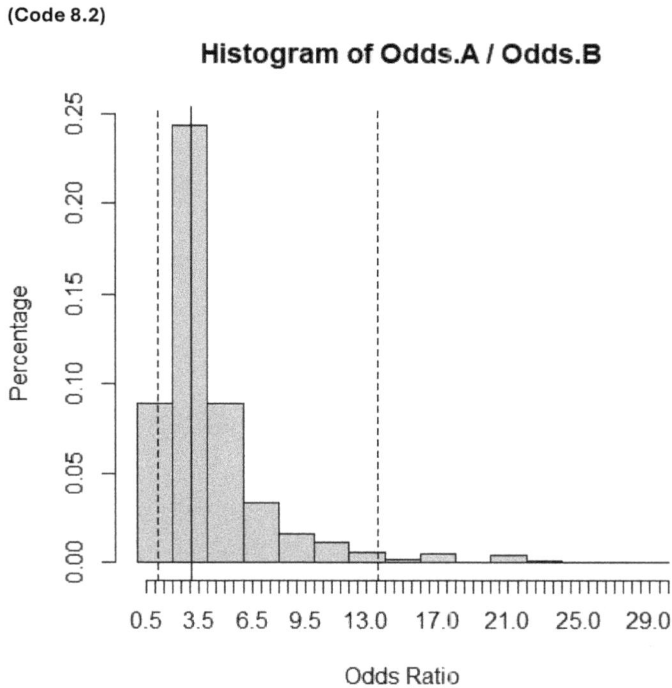

Fig. 8.4 Histogram of bootstrap OR: Odds.A/Odds.B

solid lines are the observed values from the original sample. The dashed lines are the confidence limits.

8.5 Points to Ponder

1. Must a formula exist to compute a confidence interval?
2. Do assumptions about the population distribution affect the confidence level?

References

Clopper, C. J., & Pearson, E. S. (1934). The use of confidence or fiducial limits illustrated in the case of the binomial. *Biometrika, 26*, 404–413.

Efron, B. (1982). *The jackknife, the bootstrap, and other resampling plans*. SIAM.

Scholtz, F., (2019) Confidence bounds & intervals for parameters relating to the binomial, negative binomial, poisson, and hypergeometric distributions (with Applications to Rare Events) http://faculty.washington.edu/fscholz/DATAFILES498B2008/ConfidenceBounds.pdf

Wilson, E. B. (1927). Probable inference, the law of succession, and statistical inference. *Journal of the American Statistical Association, 22*(158), 209.

Chapter 9
Linear Regression, Correlation, and Least Squares

9.1 Simple Linear Regression

Linear regression and correlation are based on having obtained a sample of paired values for two variables: (x_i, y_i): $i = 1, n$. The noticn is that the two variables are measured or observed for each of the n items/units in the sample. The notion of linearity comes from imagining a straight line tha: passes through a plot of the points (x_i, y_i). The question is which line to use?

Consider the plot in Fig. 9.1. The scatter of points might suggest that somehow a straight line is hidden within the plot. However, the question is, "Which line?"

Recall that a straight line has two "parameters," called "intercept" and "slope." The two variables are related via an equation of the form:

$$Y = \beta_0 + \beta_1 X$$

The parameters, β_0 and β_1, are called "intercept" and "slope," respectively. The slope controls the inclination of the line, which could go up (+ slope) or down (− slope). The intercept in the value of Y when $X = 0$, and graphically is the point at which the line crosses the y-axis.

In looking at Fig. 9.1, it would be hard to choose the best values for intercept and slope. Hence, the creation/discovery of least squares Least Squares is a method for finding the values of intercept and slope that would minimize the average squared vertical distance between the points in the plot and a straight line. For reasons that will be made clear later, this is called ordinary least squares (OLS) regression. Fig. 9.2 shows all the points with the least squares line added.

Supplementary Information The online version contains supplementary material available at https://doi.org/10.1007/978-3-031-84479-9_9. The videos can be accessed individually by clicking the DOI link in the accompanying figure caption or by scanning this link with the SN More Media App.

S. A. Pardo, *Beyond the T-Test*, https://doi.org/10.1007/978-3-031-84479-9_9

(Code 9.1)

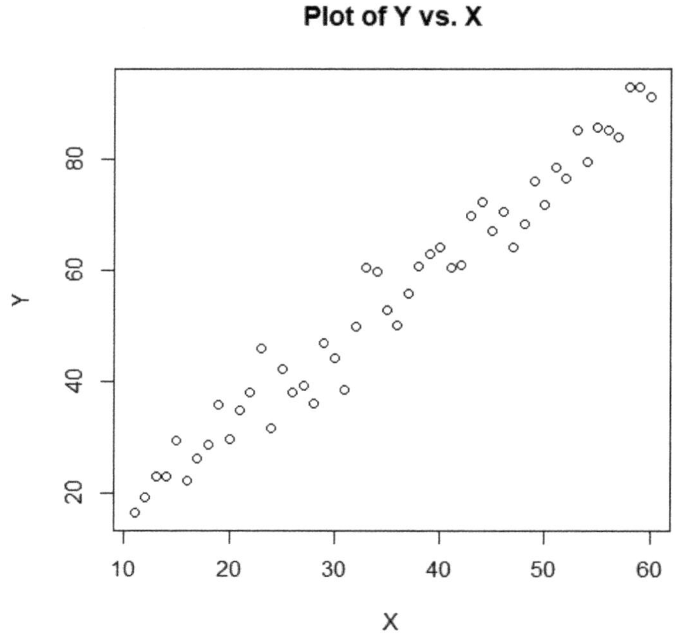

Fig. 9.1 Scatter plot of points (X, Y)

The intercept for this OLS regression line is approximately 2.3463. The slope is approximately 1.4796. In Fig. 9.2, the point at which the line touches the y-axis is approximately ($X = 0.0000$, $Y = 2.3463$).

The process of finding the least squares estimates of intercept and slope is referred to as "fitting" the linear model to the data. It is possible to obtain least squares estimates even if the linear model is entirely inappropriate. Fig. 9.3 shows a scatter plot of Y versus X where Y has no dependence on the value of X. The least squares line is also included.

Clearly the straight line can be fit to the data, even when it is inappropriate.

The notion is that we first choose a "model" that we believe describes mathematically the relationship between Y and X. For now, the model will always be linear. The assumption is that while the true relationship between Y and X is a straight line, there is some "noise" or "error" in Y that causes the values to depart somewhat from the line. It is also assumed that X has no random "noise". Hence, least squares is a method for finding the most likely line (i.e., intercept and slope) from which the Y values may vary around.

As such, the linear model for Y as a function of X includes a random noise variable:

$$Y = \beta_0 + \beta_1 X + \varepsilon$$

The random variable ε is assumed to be normally distributed with mean $\mu = 0$ and σ unknown. The random term is the random variation in the response, Y. That is, Y is a random variable, and the hypothesis is that its expected value is:

(Code 9.1)

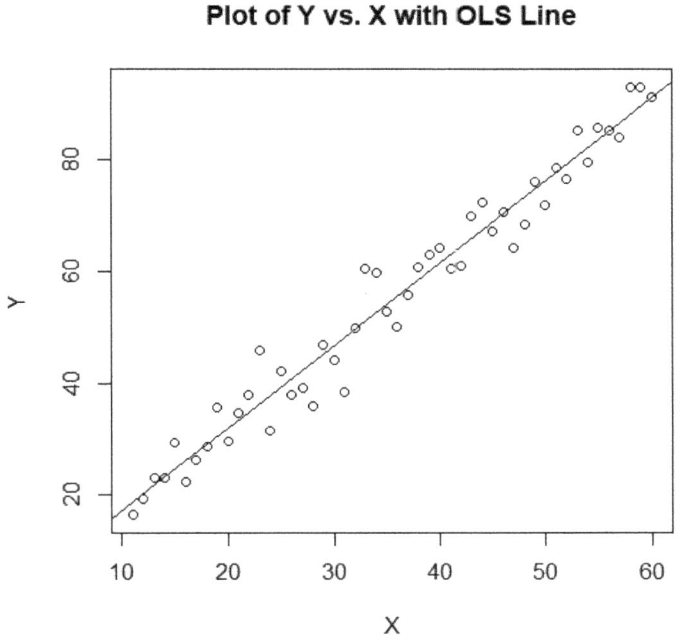

Fig. 9.2 Scatter plot with least squares line added

$$E[Y|X] = \beta_0 + \beta_1 X$$

That is, the expected value of Y depends on the value of X.

It is possible to have many variables on the right-hand side of the model equation, but for now we are restricting the discussion to having a single right-hand side variable. Such a model is called "simple" linear regression. By way of terminology, the variables on the right-hand side of the model equation (the regressors) may also be thought of as "predictors." That is we know the values of the right-hand side variables and we would like to guess at what the left-hand side, or response variable's value might be.

The question is how to assess whether the linear model is in fact appropriate. There are several measures of goodness we will employ. First we can actually test the hypotheses:

$$H_0 : \beta_i = 0, i = 0,1$$

The test statistic is:

$$t_k = \frac{\widehat{\beta}_k - 0}{SE_k}, k = 0,1$$

(Code 9.1)

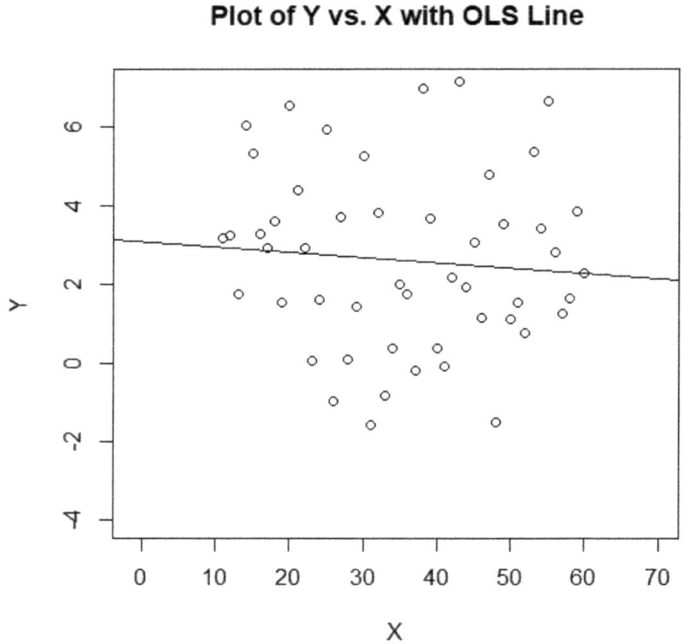

Fig. 9.3 Scatter plot of *Y* versus *X*: *Y* unrelated to *X*

The $\widehat{\beta}_k$ are the least squares estimates of the linear parameters, and SE_k are the standard errors of those estimates.

The formulas for SE_k are not too complicated.

If we define:

$$\mathrm{SS}_x = \sum_{i=1}^{n} (x_i - \overline{x})^2$$

and

$$\widehat{\sigma}^2 = \frac{1}{n-2} \sum_{i=1}^{n} \left(y_i - \widehat{\beta}_0 - \widehat{\beta}_1 x_i\right)^2$$

Then the formulas for the standard errors of the two least squares estimates are:

$$\mathrm{SE}_0 = \sqrt{\frac{1}{n} \frac{\widehat{\sigma}^2 \sum_{i=1}^{n} x_i^2}{\mathrm{SS}_x}}$$

$$\mathrm{SE}_1 = \sqrt{\frac{\widehat{\sigma}^2}{SS_x}}$$

Generally, software computed p-values for these two statistics. Another measure of goodness is called R^2 (R squared). A formula for R^2 is (Draper & Smith, 1981):

$$R^2 = 1 - \frac{\sum_{i=1}^{n} \left(\hat{y}_i - \bar{y}\right)^2}{\sum_{i=1}^{2} \left(y_i - \bar{y}\right)^2}$$

and

$$\hat{y}_i = \hat{\beta}_0 + \hat{\beta}_1 x_i$$

The $\hat{y}_i$ are the values of Y predicted by the regression equation at each value of x_i. The y_i are simply the observed values of Y at each value of X.

One shortcoming of this formula is that it does not account for the number of parameters in the model. In the case of simple linear regression, there are only 2, namely intercept and slope.

For the first least squares fit, Table 9.1 provides the statistics for the least squares estimates. To account for these, the formula is altered and is called adjusted R2. This is a formula for it:

$$R_a^2 = 1 - \left(1 - R^2\right)\left(\frac{n-1}{n-p}\right)$$

The value of p is the number of parameters in the model, which in the case of simple linear regression is 2.

In the cases of both R^2 and R_a^2, their values can be between 0 and 1. A value of 1 means there are no deviations from the least squares model at all. A value of 0 means there is no linear relationship between Y and X.

Table 9.1 Summary of linear least squares fit

(Code 9.1)

```
Coefficients:
            Estimate  Std. Error  t value   Pr(>|t|)
(Intercept) 2.34626   1.58148     1.484     0.144
X           1.47957   0.04127     35.852    <2e-16 ***
---
Signif. codes: 0 '***' 0.001 '**' 0.01 '*' 0.05 '.' 0.1 ' ' 1

Residual standard error: 4.211 on 48 degrees of freedom
Multiple R-squared: 0.964,   Adjusted R-squared: 0.9632
F-statistic: 1285 on 1 and 48 DF, p-value: < 2.2e-16
```

In this case, the p-value for the intercept was 0.144, indicating that it does not look like the estimate of 2.34626 is large enough in absolute value that we should believe the "true" or population value for the intercept is not 0. The p-value for the slope estimate of 1.47957 was less than 2×10^{-16}, which means we should reject the hypothesis that the slope is actually 0 (a horizontal line has a slope of 0). Furthermore, the adjusted R^2 was 0.9632, which is fairly close to 1. All the evidence suggests that the true relationship between Y and X is in fact linear.

Consider the least squares line shown in Fig. 9.3. Table 9.2 shows the regression summary.

In this case, the p-value for the intercept of 3.11085 was 0.000616, so we should believe that the actual intercept is not 0. However, the slope of −0.01365 had a p-value of 0.540601, so we can believe that the population slope is in fact 0. The R^2 value was close to 0 (0.007851) indicating that there is no linear relationship between Y and X. The adjusted R^2 was actually negative, which is just an artifact of computational methods. A negative value for either R^2 or adjusted R^2 can be interpreted as a 0.

The measures R^2 or adjusted R^2 are not the only or even most definitive measures of linearity. However, especially when there is only one variable on the right side of the model equation, they are adequate.

One other point about the parameter estimates and hypothesis testing. Sometimes we are interesting in knowing if the population slope is some particular number other than 0. An hypothesis test could be performed, using the test statistic:

$$t_1 = \frac{\widehat{\beta}_1 - \gamma}{SE_1}$$

Table 9.2 Summary of linear least squares fit—no linear relationship

(Code 9.1)

```
Coefficients:
             Estimate  Std. Error  t value  Pr(>|t|)
(Intercept)  3.11085   0.84860     3.666    0.000616 ***
X           -0.01365   0.02214    -0.616    0.540601
---
Signif. codes:  0 '***' 0.001 '**' 0.01 '*' 0.05 '.' 0.1 ' ' 1

Residual standard error: 2.26 on 48 degrees of freedom
Multiple R-squared:  0.007851,  Adjusted R-squared:  -0.01282
F-statistic: 0.3798 on 1 and 48 DF,  p-value: 0.5406
```

This is still a t-statistic, and as such it is easy to obtain a p-value for testing the hypothesis:

$$H_0 : \beta_1 = \gamma$$

Very often the test is for $\gamma = 1$.

The statistic t_1 is only truly distributed as Student's t provided that the error term is normally distributed with mean 0 and a standard deviation that is the same regardless of the value of X. However, the p-values are fairly robust to the assumption of normality.

9.2 Correlation

Another concept that is related to linear regression is correlation. Correlation is the degree to which two random variables are related linearly to each other. The correlation coefficient is the measure of correlation, and is expressed as:

$$r = \frac{\sum_{i=1}^{n}(x_i - \bar{x})(y_i - \bar{y})}{\sqrt{\sum_{i=1}^{n}(x_i - \bar{x})^2 \sum_{i=1}^{n}(y_i - \bar{y})^2}}$$

This formula is known as Pearson's correlation coefficient (Armitage, 1971). It has values between -1 and $+1$. The more linear the relationship, the close the correlation will be to either -1 (negative slope) or $+1$ (positive slope).

There is no specific value of the r statistic that is somehow universally meaningful, other than 0 or ± 1. However, there is a way to test hypotheses about the correlation.

Keep in mind that Pearson's r is an estimate of the population correlation:

$$\rho = \frac{E\left[(X - \mu_X)(Y - \mu_Y)\right]}{\sigma_X \sigma_Y}$$

The numerator is called the covariance of X and Y. The statistic:

$$t = \frac{r}{\sqrt{\frac{1 - r^2}{n - 2}}}$$

has a t-distribution with $n - 2$ degrees of freedom (Armitage, 1971).

The relationship between r and R^2 is fairly simple. R^2 is r^2. So r is the square root of R^2, with the sign of the slope estimate. That is:

$$r = \pm\sqrt{R^2}$$

In the first simple linear regression example $R^2 = 0.964$. So

$$r = +\sqrt{0.964} \approx 0.9818$$

The sign is + because the slope estimate was positive.
Now suppose we want to test the hypothesis:

$$H_0 : \rho = 0$$

against the alternative:

$$H_1 : \rho \neq 0$$

We first compute the t-statistic:

$$t = \frac{r}{\sqrt{\frac{1-r^2}{n-2}}} = \frac{0.9818}{\sqrt{\frac{1-(0.9818)^2}{50-2}}} \approx 35.8161$$

The p-value for this value is approximately 0.0000. So we reject H_0 in favor of H_1.
In the second example, $R^2 = 0.007851$, so

$$r = +\sqrt{0.007851} \approx 0.08861$$

Forming the t-statistic:

$$t = \frac{r}{\sqrt{\frac{1-r^2}{n-2}}} = \frac{0.08861}{\sqrt{\frac{1-(0.08861)^2}{50-2}}} \approx 0.6163$$

The p-value is approximately 0.2703, so we fail to reject H_0, and therefore believe that the population correlation, ρ, is 0.

9.3 Multiple Linear Regression

The least squares problem can be expressed in terms of matrices and vectors. Suppose the Y values are arrayed in a column vector:

$$y = \begin{bmatrix} y_1 \\ \vdots \\ y_n \end{bmatrix}$$

The X values can also be represented in the same way. However, consider that each y_i is postulated to be a linear function of the corresponding x_i, plus noise, i.e.,

$$y_i = \beta_0 + \beta_1 x_i + \varepsilon_i$$

If the errors, ε_i, are represented by the vector:

$$\boldsymbol{\varepsilon} = \begin{bmatrix} \varepsilon_1 \\ \vdots \\ \varepsilon_n \end{bmatrix}$$

Then the set of n equations could be represented in matrix form:

$$\begin{bmatrix} y_1 \\ \vdots \\ y_n \end{bmatrix} = \begin{bmatrix} 1 & x_1 \\ \vdots & \vdots \\ 1 & x_n \end{bmatrix} \begin{bmatrix} \beta_0 \\ \beta_1 \end{bmatrix} + \begin{bmatrix} \varepsilon_1 \\ \vdots \\ \varepsilon_n \end{bmatrix}$$

So representing the X part as:

$$\boldsymbol{X} = \begin{bmatrix} 1 & x_1 \\ \vdots & \vdots \\ 1 & x_n \end{bmatrix}$$

and the intercept and slope as:

$$\boldsymbol{\beta} = \begin{bmatrix} \beta_0 \\ \beta_1 \end{bmatrix}$$

the set of equations can be represented as a single matrix equation:

$$\boldsymbol{y} = \boldsymbol{X}\boldsymbol{\beta} + \boldsymbol{\varepsilon}$$

The error vector is thus:

$$\boldsymbol{\varepsilon} = \boldsymbol{y} - \boldsymbol{X}\boldsymbol{\beta}$$

The squared errors are obtained by multiplying the vectors by their transposes:

$$\boldsymbol{\varepsilon}^T \boldsymbol{\varepsilon} = \left(\boldsymbol{y} - \boldsymbol{X}\boldsymbol{\beta}\right)^T \left(\boldsymbol{y} - \boldsymbol{X}\boldsymbol{\beta}\right)$$

Simplifying, taking derivatives with respect to the parameters β_0 and β_1, setting them equal to 0 and solving gives the least squares solutions for the intercept and slope:

$$\widehat{\boldsymbol{\beta}} = \left[\boldsymbol{X}^T \boldsymbol{X}\right]^{-1} \boldsymbol{X}^T \boldsymbol{y}$$

The "hat" over the parameter vector indicates that this solution is a sample estimate, based on the data (x_i, y_i). For those readers who may not be very familiar with matrix and vector notation, see the Appendix.

There are cases where the "*Y*" variable (response)is associated simultaneously with multiple "*X*" variables (regressors). Suppose there are only two regressors, *X*1 and *X*2. Then perhaps the "model," or equation that describes the relationship between *Y*, *X*1, and *X*2 is:

$$Y = \beta_0 + \beta_1 X1 + \beta_2 X2 + \varepsilon$$

The only thing that changes to the least squares solution is the dimensionality of the ***X*** matrix and the $\boldsymbol{\beta}$ vector. That is, the *X* matrix has an additional column, and the $\boldsymbol{\beta}$ vector another row:

$$\boldsymbol{X} = \begin{bmatrix} 1 & x1_1 & x2_1 \\ \vdots & \vdots & \vdots \\ 1 & x1_n & x2_n \end{bmatrix}$$

and

$$\boldsymbol{\beta} = \begin{bmatrix} \beta_0 \\ \beta_1 \\ \beta_2 \end{bmatrix}$$

R^2, R_a^2, and *t*-tests for the parameters can all be computed for multiple regression and their interpretations are the same as in the case of simple linear regression.

One challenge in including multiple regressors is whether all of the regressors ought to be included. There are several possible methods for determining which of all potential regressors ought to be included in a model. Those methods will be described in a later chapter.

As an example of a multiple regression, suppose we suspect that systolic blood pressure is affected by a person's sodium concentration and the number of steps they take daily. The model could be expressed as:

$$Systolic = \beta_0 + \beta_1 Sodium + \beta_2 Steps + \varepsilon$$

Using least squares, the "fit" results are given in Table 9.3.

So all three parameters are significant, and R^2 and R_a^2 are both greater than 0.995. Based on these data, we would conclude that systolic blood pressure is affected by both sodium concentration and the number of daily steps, and is in fact related linearly to both regressors.

One might ask whether it would be just as effective to fit two separate models, one with sodium and the lone regressor and the other with steps by itself. Such modeling would omit the possibility of detecting an "interaction" effect; the joint

Table 9.3 Summary of multiple regression: systolic versus sodium, steps

(Code 9.2)

Coefficients:

	Estimate	Std. Error	t value	Pr(>|t|)
(Intercept)	2.257e+02	2.317e+00	97.41	<2e-16 ***
Sodium	7.353e-01	1.391e-02	52.88	<2e-16 ***
Steps	-2.854e-02	1.811e-04	-157.64	<2e-16 ***

Signif. codes: 0 '***' 0.001 '**' 0.01 '*' 0.05 '.' 0.1 ' ' 1

Residual standard error: 0.5218 on 47 degrees of freedom

Multiple R-squared: 0.9984, Adjusted R-squared: 0.9983

F-statistic: 1.424e+04 on 2 and 47 DF, p-value: < 2.2e-16

Table 9.4 Multiple regression with interaction

(Code 9.2)

Coefficients:

	Estimate	Std. Error	t value	Pr(>|t|)
(Intercept)	4.314e+01	2.119e+01	2.036	0.0475 *
Sodium	2.048e+00	1.522e-01	13.457	< 2e-16 ***
Steps	6.486e-05	3.314e-03	0.020	0.9845
Sodium:Steps	-2.056e-04	2.380e-05	-8.638	3.45e-11 ***

Signif. Codes: 0 '***' 0.001 '**' 0.01 '*' 0.05 '.' 0.1 1 ' '

Residual standard error: 0.3257 on 46 degrees of freedom

Multiple R-squared: 0.9994, Adjusted R-squared: 0.9993

F-statistic: 2.439e+04 on 3 and 46 DF, p-value: < 2.2e-16

effect of sodium and steps (e.g., perhaps taking more steps would counteract the effects of sodium levels). Consider the model:

$$Systolic = \beta_0 + \beta_1 Sodium + \beta_2 Steps + \beta_{12} Sodium : Steps + \varepsilon$$

The product of sodium and steps (*Sodium:Steps*) is referred to as an interaction effect.

Table 9.4 provides the least squares results for the model with an interaction term.

Notice that when the interaction effect was included, the Steps effect became non-significant.

We now have seen how regression can provide some insight into the relationship between some response and even multiple regressors. We have also seen how multiple regressors my "interact," and jointly impart some effect on the response. The point of this discussion is to demonstrate the concept of least squares estimation.

One other interesting fact. If the error term, ε, has a normal distribution, and its mean and variance are not affected (changed) by the values of the regressors, then the Gauss-Markov theorem (Hinkelmann & Kempthorne, 2008) states that the least squares estimates of the parameters are also the maximum likelihood estimates. The theorem was first formulated (and proven) by Carl Friedrich Gauss in the early 1800s, and then rediscovered by Andrei Markov in the early 1900's.

One other note on linearity. We mentioned earlier that linearity was based on linearity in the parameters. So, for example, the model:

$$Y = \beta_0 + \beta_1 t + \beta_2 t^2 + \varepsilon$$

is a linear model, even though one of the regressors is a squared term (t^2). That is, if we made the substitution:

$$X_1 = t$$

$$X_2 = t^2$$

then the model would become:

$$Y = \beta_0 + \beta_1 X_1 + \beta_2 X_2 + \varepsilon$$

which is linear in the parameters β_0, β_1, and β_2. Nonlinear models will be introduced later.

9.4 Some Alternatives to Least Squares for Simple Linear Regression

There are three alternatives to ordinary least squares (OLS) regression, that are particularly applied when there is only one regressor and one response. The first is often referred to as "Deming" regression (Linnet, 1990), the second called by the names of its inventors, Passing-Bablok (Passing & Bablok, 1983), and the third is weighted least squares (WLS) (Draper & Smith, 1981). Deming regression is based on the issue of having regressors that have random variation themselves (errors) as well as the response variable. Deming regression is also referred to as "errors in regressors" or "orthogonal" regression. The Passing-Bablok method is an alternative to least squares in which the slope for every pair of points is computed and then the median slope is used as the estimate. It has not been shown that there is any condition in which the Passing-Bablok method is advantageous over OLS. However, Passing-Bablok has been used especially in the validation of chemical analytic

(Code 9.2)

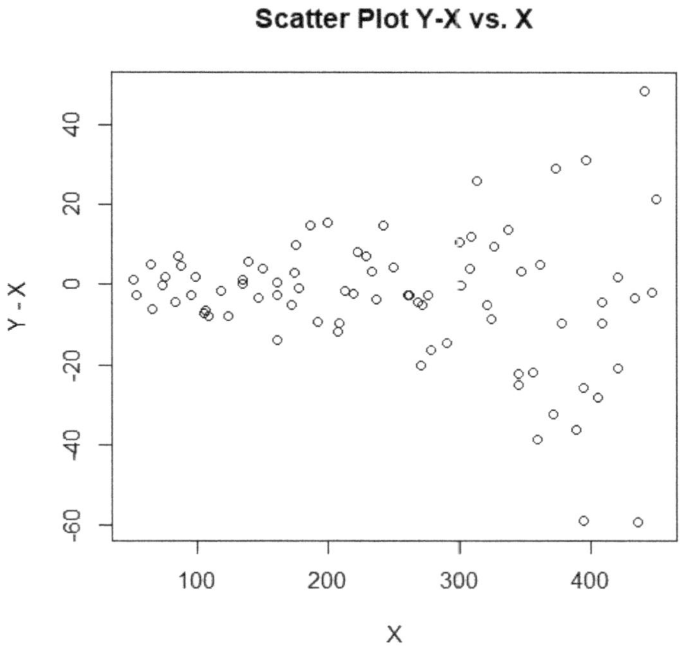

Fig. 9.4 *Y–X* when the variance in *Y* is proportional to X^{-2}

methods, including medical testing methods. WLS is useful when the variance of the response, *Y*, is proportional to the value of the regressor, *X*. That is, if:

$$V[Y|X] = V[\varepsilon] = f(X)$$

Commonly:

$$V[Y|X] = k^2 X^{-2}$$

To detect this situation, plot the difference between the response, *Y*, and the regressor, *X*, as shown in Fig. 9.4.

Notice that the variability in the difference seems to grow as the value of *X* increases. This sort of "funnel" pattern is typical when the variance in *Y* is proportional to X^{-2}.

9.5 Partial Least Squares

The least squares and multiple regression models described so far require that the number of observations meets or exceeds the number of terms in the model. Sometimes that is not the case.

One solution to these issues is called partial least squares (PLS) (Wold, 1966).

In PLS, we attempt to find a "weighting" matrix, **W**, and then compute a new "regressor" matrix:

$$\mathbf{T} = \mathbf{XW}$$

The **W** matrix is computed (we will discuss how) so that the model:

$$\mathbf{y} = \mathbf{T}\boldsymbol{q} + \varepsilon$$

can be fit using ordinary least squares. The parameter estimates, $\boldsymbol{q}$, can be related to the original regression parameters:

$$\beta = \mathbf{W}\boldsymbol{q}$$

There are multiple ways of getting the **W** matrix. One way is called Nonlinear Iterative PArtial Least Squares (NIPALS). It works like this:

```
Let M1 = XTX, A1 = XTy, C1 = Ik
For j = 1, k
 Compute qj = AjTAj
 Compute wj = CjAjqj; normalize wj by dividing by its length, and
replace it with the normalized   version;
 Store normalized wj as the jth column of a k x k matrix, W.
 Compute cj = wjTMjwj , pj = Mjwj/cj , qj = AjTwj/cj
 Store qj as the jth  row in the k × 1 vector q
 Store Aj = Aj - cjpjqjT
 Store Mj = Mj - cjpjqjT
 Store Cj = Cj - wjpjT
 Loop (next j)
```

In the end, you will have a weight matrix, **W**, that is $k \times k$, where k is the number of parameters (regressors). The PLS coefficients (parameter estimates) for the regressors would then be:

$$\widehat{\beta} = \mathbf{W}\boldsymbol{q}$$

There are other algorithms for computing the PLS estimates, but there is no reason to believe one is "better" than any other in terms of their statistical properties. Note that PLS can be easily extended to include multivariate regression, where there are multiple response variables all being predicted by the same regressors.

PLS is particularly useful when there are more regressors than observations ($k > n$).

9.6 Ridge Regression

Ridge regression (Draper & Smith, 1981) was invented to compensate for regressors that are correlated with each other. That is, some regressors are linear combinations of others. As mentioned earlier, another term for correlation between regressors is "collinearity."

Collinearity means that one is linear function of other columns in the ***X*** matrix. For example, it cannot be the case that:

$$\boldsymbol{x}_k = a_1\boldsymbol{x}_i + a_2\boldsymbol{x}_j$$

One of the consequences of having more regressors than observations is that correlation (collinearity) between regressors is induced. Thus, ridge regression and its related methods can be used to fit models when $k > n$.

In OLS, the idea is to find values of the parameters that minimize the squared "error":

$$\epsilon^2 = (y - \mathbf{X}\beta)^{\mathrm{T}}(y - \mathbf{X}\beta)$$

As we have noted earlier, this problem cannot be solved when $k > n$. In ridge regression, they add another condition to the minimization problem. Instead of finding the parameter values that minimize the squared error, ridge regression attempts to minimize:

$$\eta^2 = (y - \mathbf{X}\beta)^{\mathrm{T}}(y - \mathbf{X}\beta)$$

subject to:

$$\|\beta\|_2 \leq t$$

where t is some predetermined maximum allowable value for the "norm" $\|\boldsymbol{\beta}\|_2$.

The symbol $\|\boldsymbol{x}\|_2$ is the Euclidean length of vector x, or as it is also called, the L_2 norm of vector $\boldsymbol{x}$. So, if $\boldsymbol{\beta}$ is the vector of parameters (regression coefficients):

$$\beta = \begin{bmatrix} \beta_1 \\ \vdots \\ \beta_k \end{bmatrix}$$

then

$$\|\beta\|_2 = \left(|\beta_1|^2 + |\beta_2|^2 + \ldots + |\beta_k|^2\right)^{\frac{1}{2}}$$

It turns out that trying to minimize η^2 when some of the regressors are correlated with each other is possible. It often gives estimates that are very close to 0 for some of the parameters.

In general, the L_p norm of $\boldsymbol{\beta}$ is:

$$\|\boldsymbol{\beta}\|_p = \left(|\beta_1|^p + |\beta_2|^p + \ldots + |\beta_k|^p\right)^{\frac{1}{p}}$$

9.7 Least Absolute Shrinkage and Selection Operator (LASSO)

The name of LASSO makes it sound very special, but it isn't. It is just like ridge regression, with a slight difference:

LASSO attempts to minimize:

$$\eta^2 = (y - \mathbf{X}\beta)^{\mathrm{T}}(y - \mathbf{X}\beta)$$

subject to:

$$\|\boldsymbol{\beta}\|_1 \le t$$

That is, the constraint condition on LASSO is the L_1 norm of the parameter vector, instead of the L_2 norm. There is actually no reason to believe that LASSO is somehow better than ridge regression.

The so-called Bayesian LASSO is just LASSO under the assumption that each of the regression parameters, β_i, has a LaPlace prior distribution. In other words, Bayesian LASSO assumes that the parameters have LaPlace prior distributions with mean 0, and that they are all independent of each other (which had no relation to the independence or dependence of regressors from each other). The LaPlace distribution (also called "double exponential") has the density function:

$$f(\beta) = \frac{1}{2\theta}\exp\left(-\frac{|\beta - \mu|}{\theta}\right)$$

In the case of the Bayesian LASSO, μ is assumed to be 0, and θ is assumed to be the same for all the regression parameters.

If you were to use Bayes' theorem to compute a posterior mean for the regression parameters with the LaPlace prior assumption, you would get the same result as LASSO.

Note that in software, you might see as output the value of a parameter called lamda; this "parameter" is a product of the method by which the ridge regression or LASSO minimization proceeds. It is called LaGrange's method of undetermined

(Code 9.3)

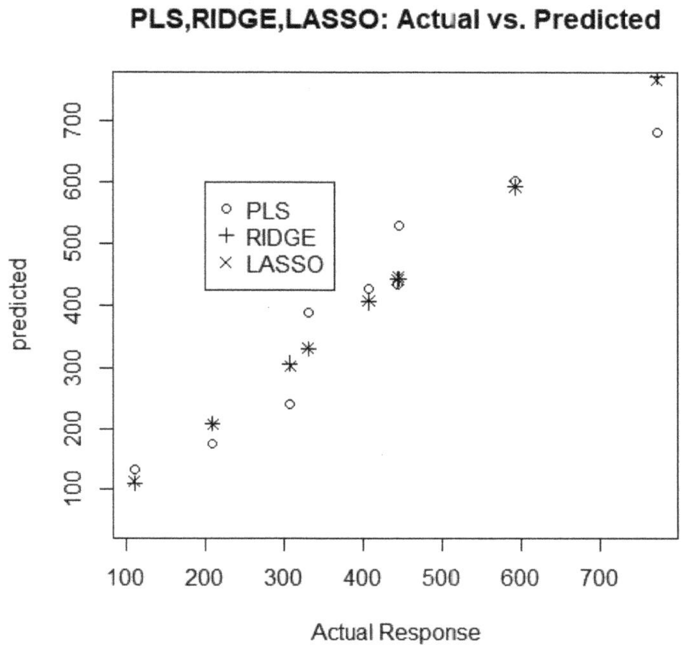

Fig. 9.5 Partial least squares (PLS), ridge regression, and LASSO predictions

multipliers. The value of lamda is the undetermined multiplier and has no meaning at all. It is just used to convert the minimization problem subject to a constraint into an easier-to-solve problem. It is of no modeling consequence.

As an example, consider a dataset with $n = 9$ observations but $k = 10$ regressors. Hence, collinearity is induced. Fig. 9.5 shows the predicted values from partial least squares, ridge regression, and LASSO regression models plotted against the observed response values.

9.8 Summary

Least squares regression is a powerful tool for finding relationships between variables. The form described in this chapter is "linear" regression, inasmuch as the equation is linear in its parameters. Simple linear regression refers to the situation where there is only one regressor. Multiple linear regression includes more than one regressor. Multiple regression may also include interaction effects with two or more of the regressors. These interactions may alter the relationship between the response variable and one or more regressors.

In addition to discovering relationships between response variables and regressors, the regression equation can be used as an interpolator, predicting values of the response for future and unobserved values of regressors.

When there are fewer observations than regressors, least squares cannot be used as such. There are methods that make transformations allowing parameter estimates to be computed. Three of those methods are partial least squares, ridge regression, and its cousin LASSO.

9.9 Points to Ponder

1. Is the model:

$$y = e^{\beta_0 + \beta_1 x_1 + \beta_2 x_2 + \varepsilon}.$$

   linear? If not, can it be linearized?
2. A multiple regression model has an adjusted R^2 (R_a^2) of 0.80. Is this good? Bad? Why?
3. Under what conditions is weighted least squares appropriate?
4. What are the basic assumptions necessary for hypothesis tests about regression parameters?

References

Armitage, P. (1971). *Statistical methods in medical research, 4th printing*. Blackwell Scientific Publications.

Draper, N. R., & Smith, H. (1981). *Applied regression analysis* (2nd ed.). John Wiley and Sons.

Hinkelmann, K., & Kempthorne, O. (2008). *Design and analysis if experiments* (Vol. 1, 2nd ed.). John Wiley and Sons.

Linnet, K. (1990). Estimation of the linear relationship between the measurements of two methods with proportional errors. *Statistics in Medicine, 9*, 1463–1473.

Passing, H., & Bablok, W. (1983). A new biometrical procedure for testing the equality of measurements from two different analytical methods. Application of linear regression procedures for method comparison studies in clinical chemistry, part I. *Journal of Clinical Chemistry and Clinical Biochemistry., 21*(11), 709–720.

Wold, H. (1966). Estimation of principal components and related models by iterative least squares. In P. R. Krishnaiaah (Ed.), *Multivariate Analysis* (pp. 391–420). Academic Press.

Chapter 10
Analysis of Variance

Analysis of Variance (ANOVA) is least squares applied to situations where the regressors have discrete values, or levels, and do not vary continuously. So, unlike sodium levels, a regressor, or "factors" as they are called when they have discrete levels, are not necessarily numeric or even ordinal. If color was a factor, its levels could be "red," "green," and "blue," for example. The levels will actually get converted into numeric codes internally by software, but the experimenter need not be concerned about such coding.

Consider a situation where there is a single "factcr" that has $k > 2$ discrete states or levels. As mentioned earlier, such an experiment is referred to as a one-way design (Montgomery, 2001). As in regression, we postulate a linear model to describe how the response variable, y, might be affected by the different levels of the "factor." The model is:

$$y_{ij} = \mu + \tau_i + \varepsilon_{ij}$$

The symbol μ represents the "overall" average value of y. The τ_i represents the average amount of difference in y from μ, when the state of the factor (experimental treatment or condition over which the observations on y are being made), is the ith level. The variable ε represents random fluctuations around the average. The double subscripts on y and ε are needed to represent the fact that for each treatment level I, there are several replicate observations made.

If y_{ij} is the jth observation of the response with factor level I, $\overline{\overline{Y}}$ is the average of all the data, regardless of factor levels, and $\overline{y}_i$ represents the average under factor level i (in our example, $i = 1, 3$), then:

Supplementary Information The online version contains supplementary material available at https://doi.org/10.1007/978-3-031-84479-9_10. The videcs can be accessed individually by clicking the DOI link in the accompanying figure caption or by scanning this link with the SN More Media App.

S. A. Pardo, *Beyond the T-Test*, https://doi.org/10.1007/978-3-031-84479-9_10

$$\text{SST} = \sum_{i=1}^{k}\sum_{j=1}^{n_i}\left(y_{ij} - \overline{\overline{Y}}\right)^2$$

$$\text{SSM} = \sum_{i=1}^{k}\left(\overline{y}_i - \overline{\overline{Y}}\right)^2$$

$$\text{SSE} = \sum_{i=1}^{k}\sum_{j=1}^{n_i}\left(y_{ij} - \overline{y}_i\right)^2$$

where k = number of factor levels (= 3 in this example) and n_i = number or replicate observations at level i (in the example $n_i = 68$ for all values of i).

SST is called the "total" sums of squares; SSM is the "model" sums of squares, and SSE the within, residual, or "error" sums of squares. It turns out that:

$$\text{SST} = \text{SSM} + \text{SSE}$$

The null hypothesis for the one-way ANOVA is:
H_0: $\tau_i = 0$ for all $i = 1, k$, so that if it is true, then:

$$y_{ij} = \mu + \varepsilon_{ij}$$

regardless of the factor level. If this null hypothesis is in fact true, then SSM should be small compared to SSE. Let $N = \sum_{i=1}^{k} n_i$ be the total sample size. It also turns out that, if the null hypothesis is true, the variables ε_{ij} are normally distributed with expected value of 0, and the variables ε_{ij} all have the same standard deviation (σ), then the ratio:

$$F = \frac{\text{SSM} / k - 1}{\text{SSE} / N - k}$$

has an F distribution with $k - 1$ degrees of freedom for the numerator and $N - k$ degrees of freedom for the denominator. Thus, a p-value can be computed to determine whether or not to believe the null hypothesis. The sums of squares divided by the associated degrees of freedom is called the "mean square." For discrete factors, the degrees of freedom for the factor is the number of levels minus 1. The total degrees of freedom for the model is $N - 1$, i.e., the total sample size minus 1. The degrees of freedom for error (residuals) is the total degrees of freedom minus the sum of the degrees of freedom for all the non-error terms in the model.

Another reason for using ANOVA is the situation where data can be classified not only by a single factor, but also by two or more such factors. If there were two such factors, the linear model could be expressed as:

$$y_{ijk} = \mu + \tau_i + \gamma_j + \varepsilon_{ijk}$$

The variable y_{ijk} is the kth replicate value of the response for level I of treatment τ and level j of treatment γ. As in the case of the single factor, the τ_i represent the average difference from overall average, μ, when the individual is exposed to level i of factor τ, and γ_j is the average difference from μ when the individual is exposed to level j of factor γ. It is possible that the difference from μ not only depends on "treatments" τ_i and γ_j, but there is also a different effect of τ_i when factor γ is at level j. Such a differential effect of one factor dependent on the level of another is in fact an interaction, and can also be included in the linear model as:

$$y_{ijk} = \mu + \tau_i + \gamma_j + \tau\gamma_{ij} + \varepsilon_{ijk}$$

In general, null hypotheses can be simultaneously tested for each of the terms in the model. The null hypotheses would have the form:

H_0: $\tau_i = 0$.
H_0: $\gamma_j = 0$.
H_0: $\tau\gamma_{ij} = 0$.

We will let q = the number of levels of factor τ and r = the number of levels for factor γ.

To help bring some intuitive focus into the two-factor model, suppose there are two factors, τ and γ, each having $q = r = 2$ levels. Suppose further that the overall (population) mean is μ = 100. Now consider the table of "effects," Table 10.1.

The average "effect" of τ, or change from the overall average μ, is called the main effect of factor τ. Similarly, the average "effect" of γ, or change from the overall average μ, is called the main effect of factor γ.

The expected value of the response, y, depends on the levels of the two factors. For example:

$$E[y_{11}] = \mu + \tau_1 + \tau\gamma_{11} = 100 + -15 + -5 = 80.0$$

Table 10.1 Some hypothetical effects in a two-factor experiment with interaction

Interaction Effect Present	
Effect	**Value**
τ_1	-15
τ_2	15
γ_1	12.5
γ_2	-12.5
$\tau\gamma_{11}$	-5
$\tau\gamma_{12}$	20
$\tau\gamma_{21}$	5
$\tau\gamma_{22}$	-20
μ =100	

However, the expected value of y when $I = 1$, which we will denote $y_{i\cdot}$, averaged over all values of j, is:

$$E\left[y_{1.}\right] = \mu + \tau_1 + \tau\gamma_{11} + \tau\gamma_{12} = 100 + -15 + -5 + \ 20 = 100.0$$

These computations are all for population parameters; hence the use of the expectation, E[.] operator. In practice, the parameters are unknown. Generally ANOVA is used to test hypotheses about the effects of factors, and then deciding on which levels of each significant factor differ from each other.

As an example, consider the problem of culturing bacteria, the workhorses of the genetic analysis world. To grow the bacteria, a small amount of them are put into a medium which provides the nourishment for them to grow. The media are formulated in many ways. In addition to the formulation, the media can be prepared in solid or liquid form (an intermediate semi-solid form is also possible, but we will ignore that for the example).

Growth can be measured by directly counting colony-forming units (CFUs) or by measuring optical density (OD), which is positively correlated with the numbers of CFUs. OD is measured by shining light at a particular wavelength (often 600 nanometers or nm) through the medium and then measuring the wave length of the light after it had come out the other end. The higher the OD, the greater the CFUs (i.e., better growth).

Suppose the experimenter wanted to know if one of two media differ in their ability to grow the bacteria. Furthermore, she wanted to know if a preparation in liquid or solid form mattered. So the linear model for the ANOVA could be expressed as:

$$\text{OD} \sim \text{MEDIUM} + \text{PREP} + \text{MEDIUM} : \text{PREP} + \varepsilon$$

Table 10.2 ANOVA for OD

(Code 10.1)

Analysis of Variance Table

Response: OD

	Df	Sum Sq	Mean Sq	F value	Pr(>F)
factor(MEDIUM)	1	0.070083	0.070083	36.002	3.406e-06 ***
factor(PREP)	2	0.312500	0.156250	80.265	2.343e-11 ***
factor(MEDIUM):factor(PREP)	2	0.051327	0.025663	13.183	0.000137 ***
Residuals	24	0.046720	0.001947		

Signif. codes: 0 '***' 0.001 '**' 0.01 '*' 0.05 '.' 0.1 ' ' 1

The *p*-values for each factor are in the column labeled Pr(>F)

Table 10.2 provides the ANOVA output.

The factors are preceded by the word "factor" which is used to indicate to software (R) that the term on the right-hand side of the equation is a discrete factor and should be treated as such.

Table 10.3 provides the mean OD at each combination of factor levels.

Finally, the means given in Table 10.3 are plotted in Fig. 10.1 with lines joining the OD means for each MEDIUM, split by the levels of PREP. Such a plot is referred to as an interaction plot (Box et al., 1978).

From these analyses, we conclude that both MEDIUM and PREP have an effect on bacteria growth, with Medium.1 showing improvement over Medium.2. The liquid PREP is better than solid. However, the combination of Medium.1 and liquid PREP is better than all other combinations.

Table 10.3 OD means for each factor level combination

(Code 10.1)

	Liquid	Solid
Medium.1	0.598	0.296
Medium.2	0.398	0.258

(Code 10.1)

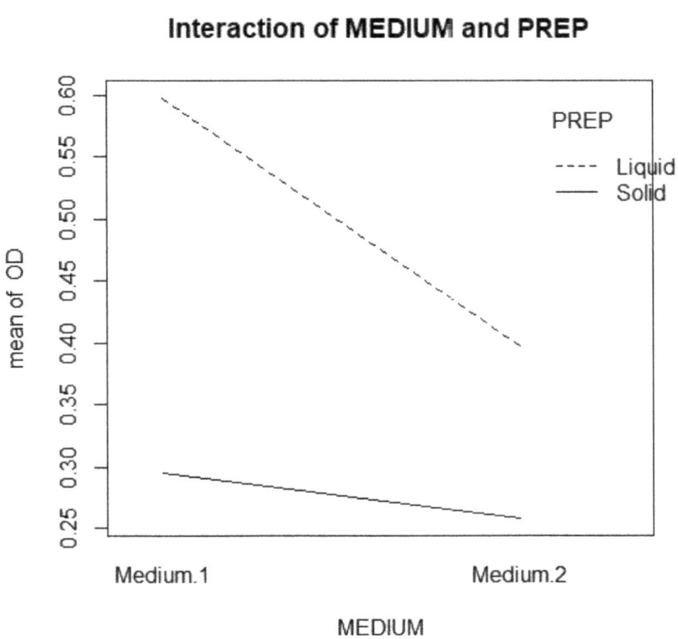

Fig. 10.1 Interaction of MEDIUM and PREP

The OD example has two factors, each having two levels/ It is possible to have more than two factors, and each factor may have different numbers of levels, with a minimum of two levels, and in theory no maximum. In practice, the number of levels is usually small, maybe even in the single digits. Furthermore, more than two factors can be involved in an interaction, although the higher the number of factors in an interaction, the more difficult the interpretation. It is generally true that the numbers of factors and levels ought to be kept as low as possible, per the principle of parsimony. Parsimony is based on the idea that the simplest explanation that encompasses most if not all of the facts is probably the best. It can be summarized by Occam's Razor:

Pluralitas non est ponenda sine necessitate

Which roughly translates to "Don't add anything (to your model) without necessity."

10.1 The Latin Square

A special experimental design is useful when each unit is subjected to all possible "treatments" or experimental devices, and the order in which those treatments are applied may have some effect on the outcome. The design family is called "Latin Square" because the treatments are symbolized by capital Latin letters, and they are arranged in a "square" indicating the order in which they are to be applied. The units (or subjects) are randomized into one group corresponding to a particular order. So, for example, if there were two treatments, labeled A and B, half the units would have A applied first and then B, and half the units would have B applied first followed by A. If there were three treatments, A, B, and C, the units would be randomized into three groups. So, one-third of the units would have the treatments applied in the order A, B, and then C; one-third would have B, C, and then A; one-third would have C, A, and then B. Table 10.4 illustrates this "rotation."

It would not matter if the groups were permuted. That is Group I could be C, A, B, Group II could be A, B, C, and Group III could be B, C, A.

The model would be:

$$Response = Group + Unit\left(Group\right) + Order + Treatment + \varepsilon$$

This brings us to a special topic: nesting. A factor is nested within another factor if its levels are different for each level of the nesting factor. So, it may be that Units

Table 10.4 A 3 × 3 Latin square design

Group	1st	2nd	3rd
I	A	B	C
II	B	C	A
III	C	A	B

Table 10.5 ANOVA for 3 × 3 Latin square experiment

(Code 10.2)

Term	Df	Sum Sq	Mean Sq	F	Pr(>F)	Code
Group	2	0	0	1.71E+00	0.186	
factor(Order)	2	0	0	6.18E-01	0.541	
Meter.Type	2	5826	2913.2	1.30E+05	2.00E-16	***
Group:factor(Subject)	48	4	0.1	3.26E+00	3.81E-07	***
Residuals	98	2	0			

Signif. Codes 0 '***' 0.001 '**' 0.01 '*' 0.05 '.'

1,4, and 7 are all in Group I, Units 2, 5, and 8 in Group II, and Units 3, 6, and 9 are in Group III. Thus, the levels of Unit depend on the level of Group.

As an example, suppose a manufacturer is comparing the accuracy of three different glucose meters, A, B, and C. They perform a study with $n = 51$ subjects, and each person tests his or her blood with each of the three meters. However, to insure that the order of testing does not interfere with assessment of the differences in accuracy, the subjects are randomized to one of three groups: A, B, C or C, A, B, or B, C, A. The response is the relative percent difference (RD) between meter reading and a laboratory instrument reading:

$$\mathrm{RD} = 100\left(\frac{Meter.Reading - Lab}{Lab}\right)$$

Table 10.5 provides the ANOVA output.

The degrees of freedom (Df) for a nested factor is the number of levels of the nesting factor, Group, multiplied by the number of levels of the nested factor within each level of the nesting factor minus 1. So, since there are 3 Groups, and 17 Subjects within each Group, the degrees of freedom for Subjects within Group are $3*(17 - 1) = 48$.

Once the treatment effect has been shown to be significant, we might want to know if all three treatments differ significantly from each other, or if two of the treatments are similar to each other and a third is different from both of them. There are tests called post-hoc multiple comparison tests, designed to answer that question. The reason for not simply performing *t*-tests for all possible pairs of treatments, or levels, is that this process will inflate the Type I error probability, i.e., the chance of concluding that two levels differ significantly when in fact they do not. One method developed by John Tukey (1949) which has become known as the honestly significant difference (HSD) test. The grandfather of all the "multiple comparison" tests is the theorem by C. E. Bonferroni (1936). The idea is that all the confidence intervals for all the pairwise differences together will have a $100\,(1 - \alpha)\%$ confidence. Table 10.6 provides the mean of the response *RD* and the results of the Tukey HSD Test.

Table 10.6 Treatment means and Tukey HSD

(Code 10.2)

Mean RD (%) BY Meter.Type		
A	B	C
-1.56	12.23	*-0.04*

Comparison	diff	lwr	upr	adjusted p-value
B-A	13.78	13.71	13.85	0.0000
C-A	1.52	1.45	1.59	0.0000
C-B	-12.26	-12.33	-12.19	0.0000

(lwr,upr) ≈ 98.33% confidence interval for pairwise differences in RD

10.2 Summary

The *t*-test was designed to compare two groups or two treatments. But what if there are more than two? Analysis of Variance (ANOVA) was created to allow the experimenter to simultaneously compare more than two treatments, without needing to do all possible pairs of treatments. In addition, there are "post-hoc" tests that, once the ANOVA shows a factor to be significant, compares the different pairs of levels while adjusting for the multiple comparisons, to avoid inflating the Type I error probability. One such method is Tukey's HSD.

A special class of experiments called Latin Squares are useful when every experimental unit is subjected to each treatment or factor level and the order in which these are applied might actually have some effect on the response. If the factor has k levels, then the experimental units are split into k groups, so that each group has N/k units (at least approximately. N is the number of units, and the number of units in any group must be an integer). The assignment is done without bias, i.e., via randomization, so that each group has a (more or less) equal number of units. One way to achieve this randomization is to assign a random order number for each on the units. Then take the first k units and assign them to Group I, then next k to Group II, etc.

10.3 Points to Ponder

1. What are the similarities between regression and ANOVA? What are the differences?
2. What is an interaction effect? How does it alter inferences?
3. If there are $N = 299$ experimental units, and the factor of interest has four levels, how many experimental units would be randomized to each group?

References

Bonferroni, C. E. (1936). Teoria statistica delle classi e calcolo delle probabilità, Pubblicazioni del R Istituto Superiore di Scienze Economiche e Commerciali di Firenze.

Box, G. E. P., Hunter, W. G., & Hunter, J. S. (1978). *Statistics for experimenters*. John Wiley and Sons.

Montgomery, D. C. (2001). *Design and analysis of experiments*. John Wiley and Sons.

Tukey, J. (1949). Comparing individual means in the analysis of variance. *Biometrics, 5*(2), 99–114.

Chapter 11
Poisson and Exponential Variables, Rate, and Time-to-Occurrence

11.1 Poisson Distribution and Rate

A rate is the number of occurrences of some event or phenomenon per unit time or space, or the number of . The event could be the passage of particles through a semi-permeable membrane per minute or it could be the number of grubs per square foot of a wheat field. The key word is "per," in that it has to do with occurrences relative to some fixed time interval or space. Rate is closely related to speed, e.g., miles per hour.

A Poisson random variable is based on the number of occurrences of an event/phenomenon per unit time (Grimmett & Stirzaker, 2004). Its basic assumption is that the average rate of occurrence does not change over time or space. Its probability mass function (discrete density) is:

$$\Pr\left\{X = x|\lambda\right\} = \frac{\lambda^x e^{-\lambda}}{x!}, x = 0,1,2,\ldots$$

The parameter is the rate in occurrences per unit time. So, if X = number of defect parts produced per day, and the average rate is $\lambda_0 = 1$ per day, then:

$$\Pr\left\{X = 1|\lambda = 1\right\} = \frac{1^1 e^{-1}}{1!} \approx 0.3679$$

The cumulative distribution for Poisson is given by a sum:

$$\Pr\left\{X \le x|\lambda\right\} = \sum_{k=0}^{x} \frac{\lambda^k e^{-\lambda}}{k!}$$

Supplementary Information The online version contains supplementary material available at https://doi.org/10.1007/978-3-031-84479-9_11. The videos can be accessed individually by clicking the DOI link in the accompanying figure caption or by scanning this link with the SN More Media App.

S. A. Pardo, *Beyond the T-Test*, https://doi.org/10.1007/978-3-031-84479-9_11

Thus the probability of having no more than 1 occurrence per unit time is:

$$\Pr\{X \le 1|\lambda\} = \sum_{k=0}^{1} \frac{\lambda^k e^{-\lambda}}{k!} = \frac{\lambda^0 e^{-\lambda}}{0!} + \frac{\lambda^1 e^{-\lambda}}{1!}$$

$0! = 1! = 1$, $\lambda^0 = 1$ and $\lambda^1 = \lambda$.

Generally the rate parameter is unknown, but it can be estimated easily. If x_1, x_2, …,x_n is a sample of the number of occurrence per unit time, for n times, then the maximum likelihood estimate of λ is:

$$\hat{\lambda} = \frac{1}{n}\sum_{i=1}^{n} x_i$$

An interesting fact about the Poisson distribution is that its variance is the same as its expected value, i.e.,

$$V(X) = E(X) = \lambda$$

That means its standard deviation is $\sqrt{\lambda}$ and the maximum likelihood estimate is:

$$\sqrt{\hat{\lambda}} = \sqrt{\frac{1}{n}\sum_{i=1}^{n} x_i}$$

The 100 (1 − α)% confidence interval is:

$$\hat{\lambda} \pm t_{n,1-\frac{\alpha}{2}} \sqrt{\frac{\hat{\lambda}}{n}}$$

$t_{n,1-\frac{\alpha}{2}}$ is the usual percentile of a Student's t distribution with n degrees of freedom.

It is straightforward to perform hypotheses about the rate parameter, λ. The null hypothesis could be:

$$H_0 : \lambda > \lambda_0$$

with alternative:

$$H_1 : \lambda \le \lambda_0$$

Perhaps the most simple way to compute a critical value for $\hat{\lambda}$ is to use Student's t distribution, i.e., if

$$t_{\hat{\lambda}} = \frac{\sqrt{n}\left(\hat{\lambda} - \lambda_0\right)}{\sqrt{\hat{\lambda}}}$$

has a Student's t distribution with n degrees of freedom, then the critical value is:

$$\hat{\lambda}_c = \lambda_0 + t_{n,1-\alpha}\sqrt{\frac{\lambda_0}{n}}$$

Note that this is a "one-sided" hypothesis test, in that we have said we only care if the rate, λ, exceeds some value, λ_0. Thus, as long as $\hat{\lambda}$ is less than $\hat{\lambda}_c$, we would reject H_0 in favor of H_1.

It is important to realize that the critical value, $\hat{\lambda}_c$, is greater than the null value λ_0. The quantity

$t_{n,1-\alpha}\sqrt{\frac{\lambda_0}{n}}$ accounts for sampling variation. Certainly in flipping a "fair" coin, even though the probability of heads landing up is 0.50, it is entirely possible to get "heads" 52 times after flipping the coin $n = 100$ times. So too if the actual rate is λ_0, it is entirely possible that the sample average might exceed this rate by some relatively small amount. That small amount is $t_{n,1-\alpha}\sqrt{\frac{\lambda_0}{n}}$. The larger the sample size, the smaller this amount becomes. Thus as the sample size increases, the power curve gets steeper. The power to reject H_0 is given by:

$$\Pr\left\{\hat{\lambda} > \hat{\lambda}_c \middle|, n, \lambda_a\right\} = \Pr\left\{t_n > \frac{\sqrt{n}\left(\hat{\lambda}_c - \lambda_a\right)}{\sqrt{\hat{\lambda}_c}}\right\}$$

and λ_a is an alternative value for the rate parameter λ. So, since we are only concerned in this case with rates that are too high, λ_a will be any value greater than λ_0.

As an example, consider a customer complaint rate. Suppose that we are willing to tolerate 1 complaint per week. Thus our hypotheses are:

$$H_0 : \lambda > \lambda_0 = 1/\text{ week}$$

and the alternative (which we hope is true):

$$H_1 : \lambda \leq \lambda_0$$

We plan to gather the data (customer complaints per week) for 52 weeks. We would like to construct the power curve to see just how good our sample size of 52 would be.

The critical value for λ is:

$$\hat{\lambda}_c = \lambda_0 + t_{n,1-\alpha}\sqrt{\frac{\lambda_0}{n}} \approx 1.23$$

The power curve is shown in Fig. 11.1.

(Code 11.1)

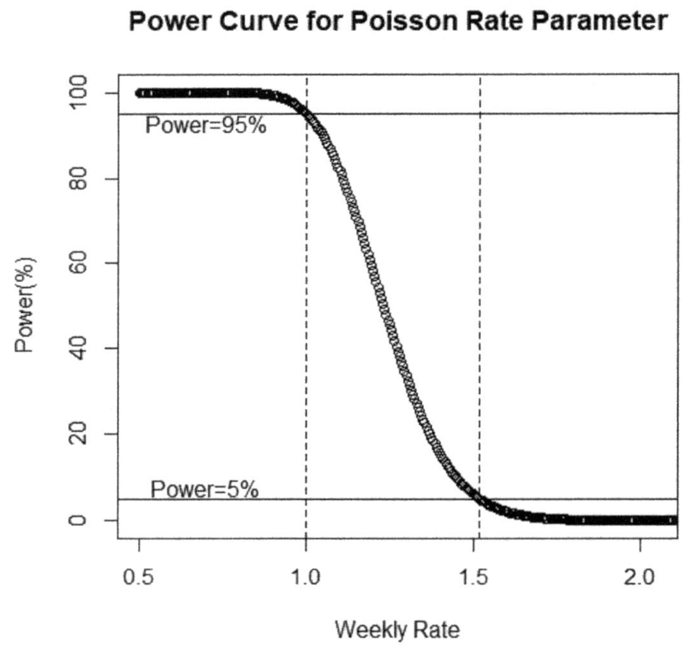

Fig. 11.1 Power curve for Poisson rate parameter

If the population rate is really 1 per week, there is a 95% chance that the sample rate estimate will be less than the critical value of 1.23. If, however, the population rate is as high as 1.52, there is only about a 5% chance that the sample rate will be less than the critical value. If you wanted to justify the use of $n = 52$, you would say that it gives you about a 95% chance of DETECTing that the population rate is as high as 1.52 per week.

11.2 Exponential Distribution

If the Poisson distribution can be used to describe probabilities associated with the number of occurrences per unit time, the exponential distribution can be used to describe probabilities associated with its reciprocal; the amount of time until the next occurrence. If the rate of failures is the number of failures/unit of time, then its reciprocal is the amount of time/next occurrence. If λ is the expected number of occurrences per unit time, then $\frac{1}{\lambda}$ is the expected amount of time until the next occurrence. Time-to-occurrence is a continuous random variable. It has a density function:

$$f(\tau|\lambda) = \tau e^{-\lambda\tau}$$

It turns out that the exponential family of random variables has a simple form for the cumulative distribution function (CDF):

$$\Pr(\mathrm{T} \le \tau|\lambda) = \int_0^{\tau} f(\varepsilon|\lambda)\,d\varepsilon = 1 - e^{-\lambda\tau}$$

and

$$\Pr\{\mathrm{T} \ge \tau|\lambda\} = e^{-\lambda\tau}$$

The expected value or average time to next occurrence is $\gamma = \frac{1}{\lambda}$ and its standard deviation is the same, $\gamma = \frac{1}{\lambda}$. Note that the exponential distribution is related to the Poisson:

$$\Pr\{X = 0|\lambda\} = \frac{\lambda^0 e^{-\lambda}}{0!} = \Pr\{\mathrm{T} \ge 1|\lambda\} = e^{-\lambda}$$

If $\tau_1, \tau_2, \ldots, \tau_n$ is a sample of times-to-occurrence, then the maximum likelihood estimate of $\gamma = \frac{1}{\lambda}$ is:

$$\hat{\gamma} = \frac{1}{\hat{\lambda}} = \frac{1}{n}\sum_{i=1}^{n}\tau_i$$

So that:

$$\hat{\lambda} = \left(\frac{1}{n}\sum_{i=1}^{n}\tau_i\right)^{-1}$$

A simple $100(1-\alpha)\%$ confidence interval for γ is:

$$\hat{\gamma} \pm t_{n,1-\frac{\alpha}{2}}\frac{\hat{\gamma}}{\sqrt{n}}$$

Note that we have expressed this as a two-sided interval (±), since it is possible that either short or long times-to-occurrence might be of interest.

The experimenter might want to test an hypothesis that there is at least $= 1 - p_0$ probability that time-to-occurrence is at least τ_0 time units, i.e.,

$$\Pr\{\mathrm{T} \ge \tau_0|\lambda_0\} \ge 1 - p_0$$

So to find the appropriate value of λ_0, solve the equation:

$$1 - p_0 = e^{-\lambda_0\tau_0}$$

to get:

$$\lambda_0 = \frac{-\ln(1-p_0)}{\tau_0}$$

Keep in mind that p_0 and τ_0 were chosen based on the requirements or specification of the experimenter.

Once λ_0 is computed (or more rarely chosen by requirements or specifications), a critical value for $\hat{\lambda}$ can be computed. Recall that $\hat{\gamma} = \frac{1}{\hat{\lambda}}$. It turns out that

$$\frac{2n\hat{\gamma}}{\gamma_0} \sim \chi^2_{2n}$$

That is, $\frac{2n\hat{\gamma}}{\gamma_0}$ has a Chi-squared distribution with $2n$ degrees of freedom.

Therefore:

$$\Pr\left\{\lambda \le \hat{\lambda} | \lambda_0\right\} = \Pr\left\{\gamma \ge \hat{\gamma} | \gamma_0\right\} = \Pr\left\{\chi^2_{2n} \ge \frac{2n\hat{\gamma}}{\gamma_0}\right\} = 1-\alpha$$

So:

$$\frac{2n\hat{\gamma}}{\gamma_0} = \chi^2_{2n,\alpha}$$

Therefore:

$$\hat{\gamma}_c = \frac{\gamma_0 \chi^2_{2n,\alpha}}{2n}$$

and

$$power = \Pr\left\{\chi^2_{2n} \ge \frac{2n\hat{\gamma}_c}{\gamma_a}\right\}, \gamma_a \ge \gamma_0 = \frac{1}{\lambda_0}$$

A power curve for testing:

$$H_0 : \lambda > \lambda_0$$

versus:

$$H : \lambda \le \lambda_0$$

(Code 11.2)

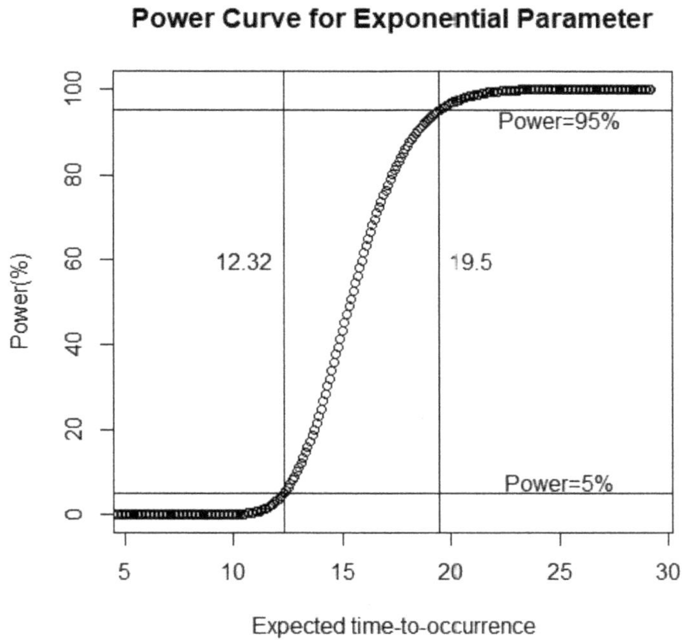

Fig. 11.2 Power curve for exponential rate (expected time-to-occurrence = 1/Rate)

With $n = 52$, $p_0 = 0.05$, $\tau_0 = 1$, $\alpha = 0.05$, the power curve is shown in Fig. 11.2. In this case,

$$\lambda_0 = \frac{-\ln(1 - p_0)}{\tau_0} \approx 0.0513$$

$$\gamma_0 = \frac{1}{\lambda_0} \approx 19.4957 \approx 19.50$$

Another way to formulate this problem is by considering the minimum sample time-to-occurrence expected if the rate parameter was λ_0.

If $\lambda = \lambda_0$, then:

$$\Pr\{T \geq \tau_0 | \lambda_0\} = 1 - p_0 = e^{-\lambda_0 \tau_0}$$

In the sample of times, τ_1, τ_2, …, τ_n, perhaps X of those times are greater than or equal to τ_0. Now we can treat X like a binomial random variable. We could compute a critical value for X, i.e., find the value x_c, such that:

$$\Pr\left\{X \geq x_c | ,n| ,1-p_0\right\} = 1-\alpha = 1-\sum_{k=0}^{x_c-1}\binom{n}{k}\left(1-p_0\right)^k p_0^{n-k}$$

So the hypotheses could be stated as:

$$H_0 : \Pr\left\{X \geq x_c\right\} < 1-p_0$$

and the alternative:

$$H_1 : \Pr\left\{X \geq x_c\right\} \geq 1-p_0$$

Thus, if the number of times is x_c or higher out of n times, H_0 is rejected in favor of H_1.

The power curve would be computed using the formula:

$$\Pr\left\{X \geq x_c | ,n| ,1-p_a\right\} = 1-\sum_{k=0}^{x_c-1}\binom{n}{k}\left(1-p_a\right)^k p_a^{n-k}$$

(Code 4.1)

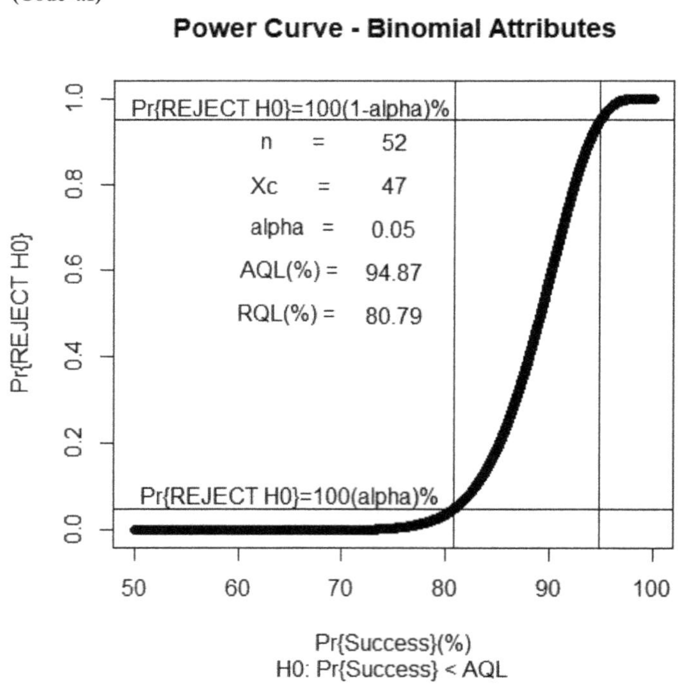

Fig. 11.3 Power curve—binomial conversion of exponential hypothesis test

for $p_a \geq p_0$. Remember that we are hoping that $\Pr\left\{T \geq \tau_0 | \hat{\lambda}_c\right\} = e^{-\hat{\lambda}_c \tau_0} = 1 - p_0$. The larger p_a gets, the smaller the probability that time-to-occurrence would be at least τ_0. With $n = 52$, $p_0 = 0.05$, alpha = 0.05, the power curve is shown in Fig. 11.3.

In this case, "success" is that a time-to-occurrence is greater than or equal to τ_0.

OK, so this is a lot to absorb. Take your time and read this over many times. The point is that time-to-occurrence is a random variable. If the rate parameter does not change with time, time-to-occurrence has an exponential distribution. This also means that the number of occurrences within a time interval of fixed length has a Poisson distribution.

11.3 More General Time-to-Occurrence Distributions

Suppose we can measure a response variable that indicates the degree to which a system is operating properly (we realize this could be challenging in many cases, but it is our point of departure.) Let $Y(t)$ be the response variable observed at time t. A first-order differential equation model that describes the change in this response over time can be expressed as:

$$\frac{dY(t)}{dt} = -\lambda Y(t)$$

With initial condition $Y(0) = y_0$, the solution to the equation is:

$$Y(t) = y_0 e^{-\lambda t}$$

Some may recognize this as the equation governing first-order chemical reaction kinetics (Whitten et al., 2004) or the equation for radioactive decay (Rutherford, 1900). The parameter λ is the rate parameter (as in the rate of reaction). If we presume that the response variable $Y(t)$ is a decreasing function of time, as in the case of analyte concentrations in chemical reactions, then it is a maximum at $t = 0$, and its maximum value is y_0. Thus we can express the response as a proportion of its initial value, namely:

$$\frac{Y(t)}{y_0} = e^{-\lambda t}$$

The proportion can be thought of at a probability, namely the probability that the value of the response is not zero after t time units. So, if we replace $Y(t)$ with a new variable, namely T = the time at which $Y(t)$ is zero, then we can express the value of $Y(t)$ as a proportion of its initial value with $\Pr\{T \geq t\}$, or:

$$\Pr\{T \geq t\} = e^{-\lambda t}$$

It turns out that, as we have already seen, this is the complement of the cumulative distribution function of an exponential random variable. That is,

$$\Pr\{T \le t\} = 1 - \Pr\{T \ge t\} = 1 - e^{-\lambda t}$$

In the language of Reliability, the parameter λ is called the failure rate. In the language of survival, it is the death rate. The exponential time-to-occurrence variable is characterized by a constant "failure" rate. That is, potentially the rate of failure could change with time, t. If $h(t)$ represents the failure rate (also called hazard rate), then for the exponential time-to-failure variable,s

$$h(t) = \lambda \forall t$$

The exponential time-to-occurrence variable is related as discussed earlier to a discrete random variable with a Poisson distribution. The variable, X, is the number of failures (or events) within a fixed length of time. It has a Poisson distribution if:

$$\Pr\{X = x|,\lambda|,t\} = \frac{(\lambda t)^x e^{-\lambda t}}{x!}$$

The value of t is a fixed length of time, and λ is the (constant) failure rate.

Suppose that the failure rate actually changes with time. For example, the rate of failure could be fairly high initially, then drop after a "burn-in" period to a constant, and then climb back up after the product reaches a "wear-out" time. Such a failure rate function is sometimes called a "bath-tub" curve, as illustrated in Fig. 11.4.

The failure rate function is also called the hazard rate function.

To generalize our initial first-order differential equation model for the response variable $Y(t)$, we could replace the constant λ with $h(t)$:

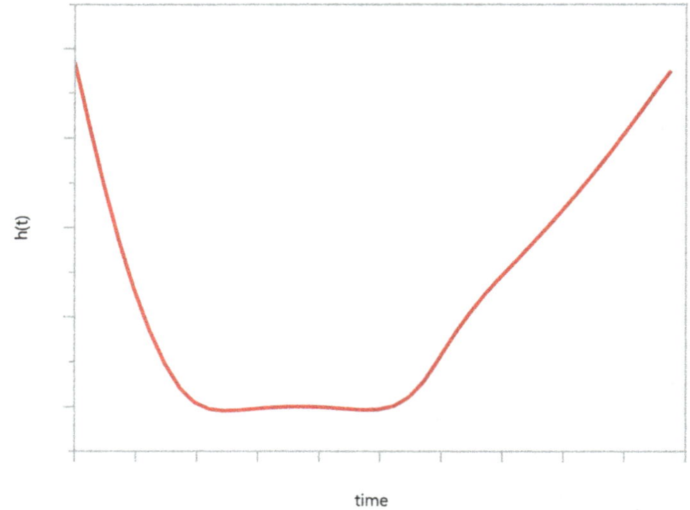

Fig. 11.4 "Bath-Tub" failure rate curve

$$\frac{dY(t)}{dt} = -h(t)Y(t)$$

Again using the initial condition $Y(0) = y_0$, the solution is

$$Y(t) = y_0 e^{-\int_0^t h(\tau)d\tau}$$

Of course, if $h(t) = \lambda$ (a constant), then:

$$\int_0^t h(\tau)d\tau = \lambda t$$

The function:

$$H(t) = \int_0^t h(\tau)d\tau$$

is called the cumulative failure rate or cumulative hazard rate function for our new random variable T, time-to-failure. So, in general, using our slightly generalized first-order kinetics model,

$$\Pr\{T \geq t\} = e^{-\int_0^t h(\tau)d\tau} = e^{-H(t)}$$

If we define the reliability function to be:

$$R(t) = \Pr\{T \geq t\} = e^{-\int_0^t h(\tau)d\tau} = e^{-H(t)}$$

Then the reliability function can be thought of as a curve in time. We also have some potentially useful relationships:

$$-\ln R(t) = H(t)$$

$$h(t) = \frac{dH(t)}{dt}$$

It is also true that $h(t)$ is related to the density function and reliability function:

$$h(t) = \frac{f(t)}{R(t)}$$

The hazard function can be interpreted as the conditional probability of occurrence in a small interval of time, $(t, t + dt)$, given that the occurrence has not happened by time t. The quantity dt is a very small amount of time, a concept from elementary calculus.

Another related pair of functions is:

$$F(t) = 1 - R(t) = 1 - e^{-H(t)}$$

$$G(t) = -\ln(1 - F(t))$$

$F(t)$ is the cumulative distribution function for time-to-occurrence.

Thus we have the relationship:

$$R(t) = 1 - e^{-G(t)}$$

The design problem is to state either the reliability function or cumulative failure rate function in terms of design parameters, and then choose values of those parameters that yield the desired reliability curve.

11.4 Obtaining an Empirical Reliability Model

Suppose the experimenter does not know the specific form of $R(t)$, $H(t)$, or $h(t)$. She or he could perform an experiment to obtain a polynomial approximation. We will call the experiment a Life Test. In this experiment, n items (devices, systems, subjects, and components) will be "started" and allowed to run until failure. The elapsed time from start to failure will be recorded. Suppose the n times, with r of those times where failure was actually observed and $n - r$ were "censored," i.e., they had not failed by the time at which they were no longer being observed. The failure times are ordered from shortest to longest. Call these times $t_1, t_2, \ldots, t_k, \ldots, t_r$. These times are referred to as "order statistics" (Conover, 1999). Compute the empirical reliability function:

$$\hat{R}(t_k) = 1 - \frac{k}{n} = \frac{n-k}{n} \quad k = 1, r$$

or the empirical cumulative distribution function:

$$\hat{F}(t_k) = 1 - \hat{R}(t_k) = \frac{k}{n} \qquad k = 1, n$$

For a lack of a better term, we will call this formula, or estimator, the empirical maximum likelihood (EML) estimator. Now suppose that the cumulative hazard rate function, $H(t)$, can be approximated by a low order polynomial in t, for example:

$$H(t) = \beta_0 + \beta_1 t + \beta_2 t^2$$

Given the relation:

$$-\ln R(t) = H(t)$$

The experimenter can now obtain via least squares an estimate of the parameters β_k, and thus an approximation formula for $R(t)$. That is, the estimated approximation formula would be:

$$\widehat{R}(t) = \exp\left(-\widehat{\beta}_0 - \widehat{\beta}_1 t - \widehat{\beta}_2 t^2\right)$$

The $\widehat{\beta}_i$ are the least-squares estimates of the $H(t)$ approximation formula parameters. The approximation formula could be used to interpolate values of $R(t)$. but interpolation is not its most important use. More importantly, it can be used as a design tool.

The mathematical explanation might seem intimidating. However, the procedure is fairly simple.

1. Arrange the sample of times-to-occurrence from smallest to largest. Call this ordered sample $t_1, t_2, \ldots, t_k, \ldots, t_n$.
2. Compute reliability/survival function:

$$\widehat{R}(t_k) = 1 - \frac{k}{n} = \frac{n-k}{n} \quad k = 1, r$$

A better estimator is:

$$\widehat{R}(t_k) = \widehat{R}(t_{k-1}) \frac{n-k}{n+1-k}, k = 1, r$$

and:

$$\widehat{R}(t_1) = \frac{r-1}{n}$$

This recursive estimator is called the Kaplan-Meier estimator (Lee, 1992). In the case where $r = n$, i.e., all n units were observed to have failed, you can either compute the reliability function up to but not including the final (rth) failure time, or modify the formulas:

$$\widehat{R}(t_k) = \widehat{R}(t_{k-1}) \frac{n+1-k}{n+2-k}$$

and:

$$\widehat{R}(t_1) = \frac{n}{n} = 1$$

The issue is that if $r = n$, then $\widehat{R}(t_r) = 0$, precluding the possibility of taking the logarithm of $\widehat{R}(t_r)$.

3. Compute empirical hazard rate function:

$$\widehat{H}(t_k) = -\ln \widehat{R}(t_k)$$

4. Fit a polynomial via least squares:

$$\widehat{H}(t_k) = -\ln \widehat{R}(t_k) = \beta_0 + \beta_1 t_k + \varepsilon$$

5. Compute p-value for the slope parameter, β_1. If it is less than α (usually 0.05), then you know that the time-to-occurrence is not exponential.
6. You might want to fit a higher-order polynomial, to get a more accurate model of the hazard function, and subsequently the Reliability/Survival function, e.g.,

Table 11.1 Results of linear fit to an empirical hazard rate function

(Code 11.3)

```
> summary(poly.1)

Call:
lm(formula = H.t ~ Time)

Residuals:
    Min      1Q  Median      3Q     Max
-0.2626 -0.1948 -0.1300  0.1002  0.8915

Coefficients:
             Estimate  Std. Error  t value   Pr(>|t|)
(Intercept) -0.984890  0.105138    -9.368    2.06e-12 ***
Time         0.104486  0.005233    19.965    < 2e-16 ***
---
Signif. codes:  0 '***' 0.001 '**' 0.01 '*' 0.05 '.' 0.1 ' ' 1

Residual standard error: 0.2945 on 48 degrees of freedom
Multiple R-squared:  0.8925,    Adjusted R-squared:  0.8903
F-statistic: 398.6 on 1 and 48 DF,  p-value: < 2.2e-16
```

Table 11.2 Results of quadratic fit to an empirical hazard rate function

(Code 11.3)

```
> summary(poly.2)

Call:
lm(formula = H.t ~ Time + tsq)

Residuals:
     Min       1Q   Median       3Q      Max
-0.48059 -0.06976 -0.04360  0.04082  0.68675

Coefficients:
              Estimate  Std. Error  t value  Pr(>|t|)
(Intercept) -0.1472293   0.1028370   -1.432     0.159
Time         0.0023772   0.0106120    0.224     0.824
tsq          0.0025913   0.0002584   10.027  2.94e-13 ***
---
Signif. codes: 0 '***' 0.001 '**' 0.01 '*' 0.05 '.' 0.1 ' ' 1

Residual standard error: 0.168 on 47 degrees of freedom
Multiple R-squared: 0.9658,  Adjusted R-squared: 0.9643
F-statistic: 662.9 on 2 and 47 DF,  p-value: < 2.2e-16
```

$$\widehat{H}\left(t_k\right) = \beta_0 + \beta_1 t_k + \beta_2 t_k^2 + \varepsilon$$

Again, a *p*-value for β_2 will tell you if the higher-order term (t_k^2) is warranted. That is mostly it.

Consider an example of time-to-occurrence data. An empirical reliability/survival function was computed. In addition, three polynomials were fit to the data: linear, quadratic, and cubic. Tables 11.1,11.2, and 11.3 provides the results of the least -squares fits.

The regressors tsq and tcb are computed as:

tsq = Time^2
tcb = Time^3

Figure 11.5 shows the empirical reliability function together with the three fitted polynomials. Which one fits the best?

Table 11.3 Results of cubic fit to an empirical hazard rate function

(Code 11.3)

```
> summary(poly.3)

Call:
lm(formula = H.t ~ Time + tsq + tcb)

Residuals:
     Min       1Q   Median       3Q      Max
-0.28482 -0.05869 -0.00546  0.03573  0.60030

Coefficients:
              Estimate  Std. Error  t value   Pr(>|t|)
(Intercept)  2.518e-01   1.221e-01   2.062    0.044899 *
Time        -8.639e-02   2.123e-02  -4.070    0.000183 ***
tsq          7.678e-03   1.126e-03   6.820    1.70e-08 ***
tcb         -8.193e-05   1.780e-05  -4.604    3.28e-05 ***
---
Signif. codes:  0 '***' 0.001 '**' 0.01 '*' 0.05 '.' 0.1 ' ' 1

Residual standard error: 0.1405 on 46 degrees of freedom
Multiple R-squared:  0.9766,    Adjusted R-squared:  0.975
F-statistic: 638.8 on 3 and 46 DF,  p-value: < 2.2e-16
```

(Code 11.3)

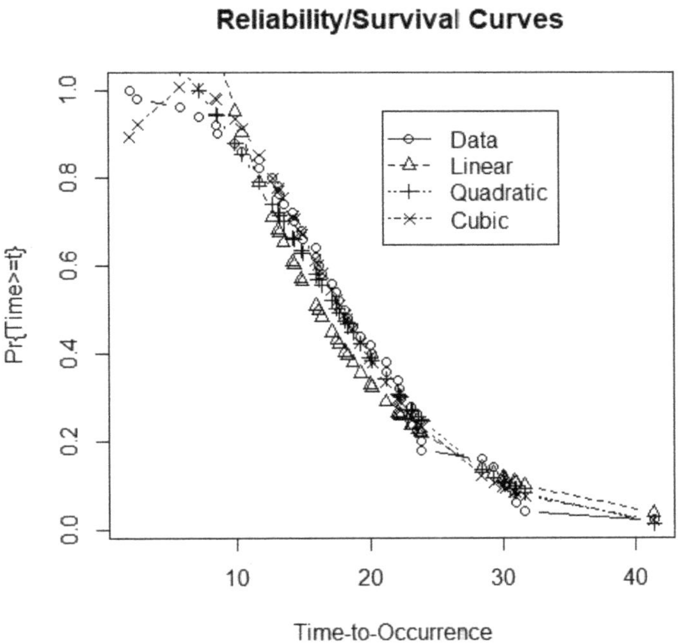

Fig. 11.5 Empirical reliability function and polynomial fits

If you would ask the question, "I now believe the rate is not a constant, but actually varies in time. What if that variation depends on other conditions, such as materials, seasons, locations, etc.?" To answer this question a factorial experiment may be able to provide some insights.

11.5 Accelerated Life Tests

Sometimes products will last for a very long time under normal use conditions. Thus, trying to estimate the probability that they will last to some warranted time may be challenging. However, many such products will fail at a much greater rate if the conditions to which they are exposed are much harsher than usual. The problem is to determine how much the harsher conditions have accelerated their failure rate. Such conditions are known as accelerated conditions, and the acceptance test under accelerated conditions is called an accelerated life test (Mann et al., 1974).

Commonly, temperature is used as the accelerating condition. Presuming that the increase in failure rate is proportional to an increase in the rate of a chemical reaction, a model called the Arrhenius Reaction Rate Law (Mann et al. 1974) is employed to relate the failure rate to temperature. The Arrhenius model is:

$$\lambda_P = A\exp\left(\frac{-E/K}{P}\right)$$

The constant A is specific to the particular materials and reactions that underlie the failure mode. The constant E is called the energy of activation, and is also specific to the materials and chemical reactions involved in failure. The letter K stands for Boltzmann's constant, and P is the temperature in degrees Kelvin (we use the letter P, for "parameter," so as to not confuse it with T for time-to-failure). So λ_P is the average failure rate at temperature P. The usual presumption is that the time-to-failure follows an exponential distribution, so that at temperature P, the reliability function is given by:

$$R(t|P) = e^{-\lambda_P t}$$

The question is how to determine the degree of acceleration achieved by exposing experimental units to a particular temperature, say P_a. The first problem is to estimate the parameters A and $B = -E/K$ (note that B is just a normalized energy of activation). The answer depends on the nature of the data. Life tests may be performed by placing n units into a temperature chamber, at a fixed temperature, P_a, for a given time, T. At time T, the units are taken out and inspected or tested, and the number of units that "survive," $S = s$, are counted. Assuming that the time-to-failure is exponentially distributed, and an estimate of the reliability at time T is $p_A = \frac{s}{n}$, then the reliability is given by:

$$\hat{R}(T|P_a) = e^{-\lambda_a T} = p_a = \frac{s}{n}$$

Solving for λ_a gives an estimate for the failure rate:

$$\hat{\lambda}_a = \frac{-\ln(p_a)}{T} = \frac{-\ln\left(\frac{s}{n}\right)}{T}$$

Suppose an experiment was performed where n_1 units were put on test for T time units at temperature P_1 and another n_2 units put on test for T time units at temperature P_2. Then we would have two equations in the two unknowns A and B:

$$-\ln\left(\frac{s_1}{n_1}\right) = TA\exp\left(\frac{-B}{P_1}\right)$$

$$-\ln\left(\frac{s_2}{n_2}\right) = TA\exp\left(\frac{-B}{P_2}\right)$$

Taking logs on both sides yields:

$$\ln\left(-\ln\left(\frac{s_1}{n_1}\right)\right) = lnT + lnA - \frac{B}{P_1}$$

$$\ln(-\ln\left(\frac{s_2}{n_2}\right) = \ln T + \ln A - \frac{B}{P_2}$$

These in turn yield the solutions for the estimates:

$$\hat{B} = P_1\left[\ln\hat{A} - \ln\left(-\ln\frac{s_1}{n_1}\right) - \ln T\right]$$

$$\ln\hat{A} = \frac{P_2}{P_2 - P_1}\left[\ln\left(-\ln\frac{s_2}{n_2}\right) + \frac{P_1}{P_2}\ln\left(-\ln\frac{s_1}{n_1}\right) + \frac{P_1}{P_2}\ln T\right]$$

Thus, an expression for the estimate of parameter B in only known quantities is:

$$\hat{B} = P_1\left[\frac{P_2}{P_2 - P_1}\left[\ln\left(-\ln\frac{s_2}{n_2}\right) + \frac{P_1}{P_2}\ln\left(-\ln\frac{s_1}{n_1}\right) + \frac{P_1}{P_2}\ln T\right] - \ln\left(-\ln\frac{s_1}{n_1}\right) + \ln T\right]$$

These estimates will allow the experimenter to determine how much acceleration was achieved at any given temperature, P_2, compared to a lower temperature, P_1. The estimates of B and A may be useful for future experiments or tests, provided that the materials involved in such tests are at least similar if not identical to those used to obtain the estimates.

Suppose that the experimenter has at least an hypothetical failure rate desired at say 25 $°C = 25 + 273 = 298°K = P_0$. Such a failure rate might be determined by having a specification or requirement that the reliability at time T_0 must be at least r_0. Assuming the exponential time-to-failure, the failure rate is given by:

$$\lambda_0 = \frac{-\ln(r_0)}{T_0}$$

Now suppose that the experimenter wants to accelerate the failure process k times, so that the actual test time would need to be $T_a = \frac{T_0}{k}$ ("a" stands for "accelerated"). This would also mean that the failure rate under the accelerated conditions would need to be:

$$\lambda_a = k\lambda_0$$

The experimenter must choose a temperature, $P_a > P_0$, to achieve the desired acceleration.

Using the Arrhenius equation:

$$k = \frac{\exp(-B/P_a)}{\exp(-B/P_0)} = \exp\left(B\left(\frac{1}{P_0} - \frac{1}{P_a}\right)\right)$$

Since k is actually given (i.e., the desired acceleration to allow the test to occur in a short enough time), and P_0 is known, the equation can be solved for P_a:

$$P_a = \frac{BP_0}{B - P_0 \ln k}$$

The only thing required is a value for B, the normalized energy of activation. A simple experiment as described earlier can be used to obtain an estimate of B.

Recall that test temperatures should be expressed in degrees Kelvin (°K) when using these equations.

There are two potential drawbacks to the procedures described for determining parameters of an accelerated life test. First, we have assumed that the time-to-failure is affected by temperature in the way described by the Arrhenius equation. Secondly, we have assumed that time-to-failure has an exponential distribution. Both of these assumptions stem from a more fundamental assumption that the failure process is related to a first-order chemical reaction (Chow, 2007). While these assumptions may not be completely valid, they may provide at least a practical approach to determining the amount of acceleration achieved by putting units on test at temperature P_a for time T_0.

As an example, suppose $n = 30$ units were put on test for 336 h (14 days) at 25 °C, and $n = 30$ units put on test at 40 °C for 336 h. At 25 °C, $s_1 = 27$ units "survived," i.e., were operating properly, and at 40 °C, $s_2 = 23$ were operating properly. Then the "normalized" activation energy is $B = 14910.26$, and to achieve an acceleration of $k = 26$ times, we would expose the units to 45.76 °C. To simulate 2 years at 25 °C, we would expose the units to 45.76 °C for 28 days. If we wanted a 99% chance of units operating properly after 2 years at 25 °C, then:

$$\lambda_0 = \frac{-\ln(0.99)}{730.5}$$

$$\lambda_a = 26\lambda_0$$

and

$$\Pr\{T_a \geq 28 | \lambda_a\} = e^{-\lambda_a 28} \approx 0.9900$$

A sampling plan with $n = 50$ and a critical number of "successes" of $x_c = 49$ would yield the power curve shown in Fig. 11.6.

(Code 4.1)

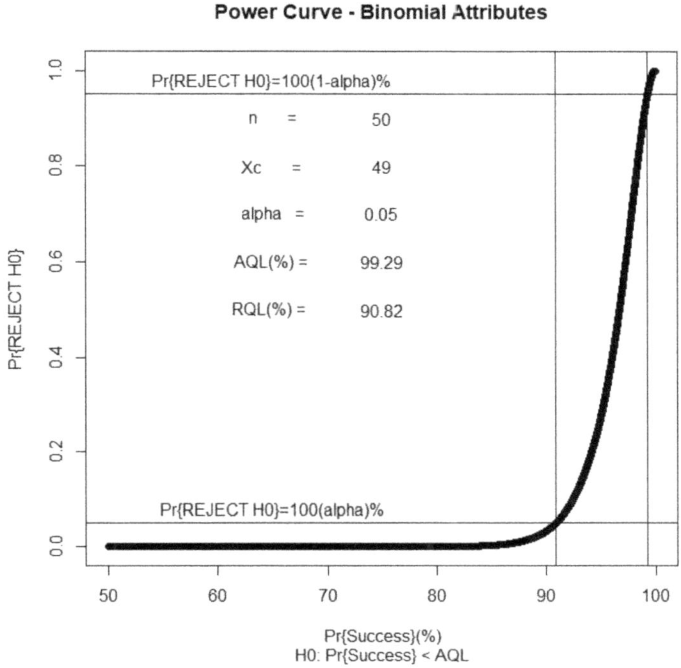

Fig. 11.6 Power curve for accelerated life test, 28.1 days at 45.76 °C

11.6 Summary

Time-to-occurrence variables are often characterized by their hazard rate function or cumulative hazard rate function. It is not that other random variables do not have any hazard rate or cumulative hazard rate functions associated with them, but these functions have a special interpretation for time-to-occurrence variables. That is, the hazard rate function represents the conditional probability of the next occurrence happening in the interval $(t, t + dt)$, given that it has not happened for t time units since the last occurrence.

Perhaps the most ubiquitous family of time-to-occurrence variables is the exponential family, which has constant hazard rates. It is possible to perform hypothesis tests for the rate parameter as well as computing confidence intervals for them.

One use of the exponential family is to generate accelerated life tests. Items that under normal conditions last a very long time can have their failure process accelerated, usually by exposing them to higher than normal temperatures. The Arrhenius reaction rate law can be used to design tests where the items are exposed to "accelerating" conditions and computing by how much the failure rate has been accelerated.

11.7 Points to Ponder

1. What is the fundamental assumption necessary for a random variable, time-to-occurrence, to have an exponential distribution?
2. Suppose the average rate of cavalry soldiers being kicked in the head by their horse was 0.5 per month. What is the probability that no soldiers will be kicked in the coming month? What is the probability that the time to the next soldier getting kicked will be greater than 1 month?
3. What are the assumptions necessary to make use of the Arrhenius Law to assess the acceleration of failure rate?

References

Chow, S.-C. (2007). *Statistical design and analysis of stability studies*. Chapman and Hall/CRC.

Conover, W. J. (1999). *Practical nonparametric statistics* (3rd ed.). John Wiley and Sons.

Grimmett, G., & Stirzaker, D. (2004). *Introduction to probability and random processes* (3rd ed.). Oxford University Press.

Lee, E. T. (1992). *Statistical methods for survival data analysis*. John Wiley and Sons.

Mann, N. R., Schafer, R. E., & Singpurwalla, N. D. (1974). *Methods for statistical analysis of reliability and life data*. John Wiley and Sons.

Rutherford, E. (1900). A radioactive substance emitted from thorium compounds. *Phil. Mag Ser., 5*, 1–14.

Whitten, K. W., Davis, R. E., Peck, M. L., & Stanley, G. G. (2004). *General chemistry* (7th ed.). Brooks/Cole.

Chapter 12
2^k Factorial Experiments

We have seen how ANOVA can be used to simultaneously assess the effects of multiple factors on a response variable. Often product or process design decisions about multiple factors must be made. One methodology is the use of two levels of each factor in an experiment that allows the simultaneous assessment of multiple factors and their interactions. That is, it is possible that each factor may have a continuum of possible values. Not only must the decision-maker determine "optimal" values for each factor, but in fact she or he must determine whether any or all of the factors actually have an effect on the response.

Inasmuch as we are speaking about continuously-valued factors, we certainly have no way to test all possible values of each factor. The idea of the 2^k experiment is that there are k continuously-valued factors which may or may not actually affect the values of the response. The experimenter will choose two levels for each of the k factors, one low value and one high value, and then perform the experiment with all 2^k combinations.

By way of an example, suppose a chemical process can be run under different values of temperature, pressure, and time. Furthermore, each of those $k = 3$ factors are continuous, and have minimum and maximum levels in which they can operate. Table 12.1 provides the values.

Table 12.1 2-level factors

Factor	Units	Minimum	Maximum
Temperature	°C	40	60
Pressure	kP (kiloPascals)	10	14
Time	Minutes	30	45

Supplementary Information The online version contains supplementary material available at https://doi.org/10.1007/978-3-031-84479-9_12. The videos can be accessed individually by clicking the DOI link in the accompanying figure caption or by scanning this link with the SN More Media App.

S. A. Pardo, *Beyond the T-Test*, https://doi.org/10.1007/978-3-031-84479-9_12

Table 12.2 2^3 runs for the 3-factor experiment

Run	Temperature	Pressure	Time
1	40	10	30
2	40	10	45
3	60	10	30
4	60	10	45
5	40	14	30
6	40	14	45
7	60	14	30
8	60	14	45

Since there are three factors, there are $2^3 = 8$ different combinations to try. Table 12.2 provides the different combinations, referred to as runs. These are the different conditions under which the process will be run.

Suppose that the response variable to be measured is Yield, the amount of the chemical product after a run is completed. Suppose further that each run is completed twice.

We can use ANOVA to assess the model:

$$\begin{aligned}\text{Yield} \sim \text{Temp} + \text{Pres} + \text{Time} + \text{Temp} : \text{Pres} + \text{Temp} : \text{Time} + \text{Pres}\\ : \text{Time} + \text{Temp} : \text{Pres} : \text{Time} + \varepsilon\end{aligned}$$

Each term in the model that has only one factor name is referred to as a "main" effect.

The notation $X_1{:}X_2$ means the interaction between factors X_1 and X_2. Likewise, $X_1{:}X_2{:}X_3$ means the interaction of all three factors X_1, X_2, and X_3.

Table 12.3 provides the ANOVA results.

The column labeled Pr(>|F) has the p-values for each term in the model. The three 2-factor interaction effects were significant; the 3-factor interaction was not.

The ANOVA provided a classification model, as opposed to a predictive model. Since the factors are all continuous, a multiple regression model can also be generated with these data.

Table 12.4 provides the results of the multiple regression. The estimates for each term are different that when the factors were treated as discrete in the ANOVA.

In the regression, the only term with a coefficient (slope) that was significantly greater than 0 was the Temp:Pres interaction. However, compare the observed Yield values with those predicted by the regression model, in Fig. 12.1. The straight line is the linear regression of Predicted against Observed Yield. This line indicates that while the model may have relatively small coefficients, they seem to help predicting Yield.

We might be interested in the results of a "reduced" model, with only Temp, Pres, and Temp:Pres terms. Table 12.5 provides the results of such a model.

In the reduced model, the main effects of Temp and Pres still do not have coefficients that differ significantly from 0. The Predicted versus Actual plot (and regression line) are shown in Fig. 12.2. The points seem to fall off the regression line with

Table 12.3 Yield experiment ANOVA—3 factors

(Code 12.1)

Analysis of Variance Table

Response: Yield

	Df	Sum Sq	Mean Sq	F value	Pr(>F)
fTemp	1	12907.8	12907.8	6787.7960	5.254e-13 ***
fPres	1	7620.0	7620.0	4007.1021	4.313e-12 ***
fTime	1	77.9	77.9	40.9781	0.0002088 ***
fTemp:fPres	1	362.8	362.8	190.7886	7.291e-07 ***
fTemp:fTime	1	17.9	17.9	9.3982	0.0154497 *
fPres:fTime	1	17.4	17.4	9.1333	0.0165058 *
fTemp:fPres:fTime	1	0.7	0.7	0.3733	0.5581855
Residuals	8	15.2	1.9		

Signif. codes: 0 '***' 0.001 '**' 0.01 '*' 0.05 '.' 0.1 ' ' 1

larger differences than in the "full" model. However, the rule of parsimony (remember Occam's Razor) indicates that perhaps this prediction is not only good enough, but that adding extra terms to the model might mean that we are "overfitting," which is something like connecting the dots for the current data. So, while the full model may appear better than the reduced model, perhaps it would not be as good with a new set of data.

Since both models have high values for adjusted R^2, it is hard to judge.

Another measure to indicate how good the model fits the data Mallow's *Cp* (Draper and Smith, 1981)

If we call the following formula the residual sums of squares (RSS):

$$\text{RSS} = \sum_{i=1}^{n} \left(\hat{y}_i - y_i\right)^2$$

Then *Cp* is defined as:

$$Cp = \frac{\text{RSS}}{\hat{\sigma}^2} - (n - 2p)$$

and $\hat{\sigma}$ in the estimate of the standard deviation of the full model, i.e., the model with all terms included.

Table 12.4 Multiple regression output for yield

(Code 12.1)

```
Call:
lm(formula = Yield ~ Temp + Pres + Time + Temp:Pres + Temp:Time +
   Pres:Time + Temp:Pres:Time)

Residuals:
  Min    1Q Median    3Q   Max
-1.69 -0.59   0.00  0.59  1.69

Coefficients:
                Estimate   Std. Error  t value  Pr(>|t|)
(Intercept)     90.812500  54.522523   1.666    0.1344
Temp            -0.120250   1.069275  -0.112    0.9132
Pres            -1.021250   4.481724  -0.228    0.8255
Time             0.989833   1.425699   0.694    0.5072
Temp:Pres        0.290750   0.087894   3.308    0.0107 *
Temp:Time        0.002758   0.027960   0.099    0.9238
Pres:Time        0.000750   0.117192   0.006    0.9951
Temp:Pres:Time  -0.001404   0.002298  -0.611    0.5582
---
Signif. codes:  0 '***' 0.001 '**' 0.01 '*' 0.05 '.' 0.1 ' ' 1

Residual standard error: 1.379 on 8 degrees of freedom
Multiple R-squared:  0.9993,   Adjusted R-squared:  0.9986
F-statistic:  1578 on 7 and 8 DF,  p-value: 6.428e-12
```

It turns out that if in fact the true relationship between Y and X has p parameters, then the expected value of Cp is approximately p. This is a way of deciding whether the model is at least correct in the number of parameters.

The Cp for the full model will always be exactly the number of parameters. For the reduced model, it turned out $Cp \approx 59.9$, which is certainly higher than 8. If the Time factor is added back into the model, $Cp \approx 20.9$, which is certainly a great improvement.

(Code 12.1)

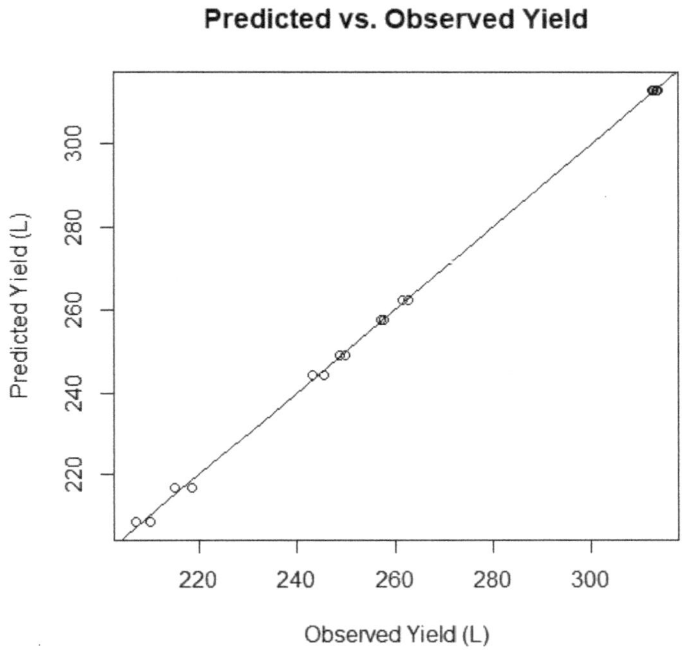

Fig. 12.1 Predicted versus observed yields

Model choice is more of an art than science. There are no absolute rules about which model is best. We have seen several tools for helping to decide on which model to rely: significance tests, R^2, Cp, and plots of predicted versus actual response values. There are some procedures to help use these tools in an orderly if not completely objective fashion. Those procedures will be discussed in Chap. 15.

Table 12.5 Multiple regression—reduced model

(Code 12.2)

```
Call:
lm(formula = Yield ~ Temp + Pres + Temp:Pres)

Residuals:
    Min      1Q  Median      3Q     Max
-5.5750 -2.2944 -0.0175  1.9681  5.7850

Coefficients:
              Estimate  Std. Error  t value  Pr(>|t|)
(Intercept)  127.93125    25.43190    5.030  0.000294 ***
Temp          -0.01681     0.49876   -0.034  0.973664
Pres          -0.99312     2.09049   -0.475  0.643268
Temp:Pres      0.23809     0.04100    5.807  8.38e-05 ***
---
Signif. codes:  0 '***' 0.001 '**' 0.01 '*' 0.05 '.' 0.1 ' ' 1

Residual standard error: 3.28 on 12 degrees of freedom
Multiple R-squared:  0.9939,    Adjusted R-squared:  0.9923
F-statistic: 647.3 on 3 and 12 DF,  p-value: 1.569e-13
```

(Code 12.2)

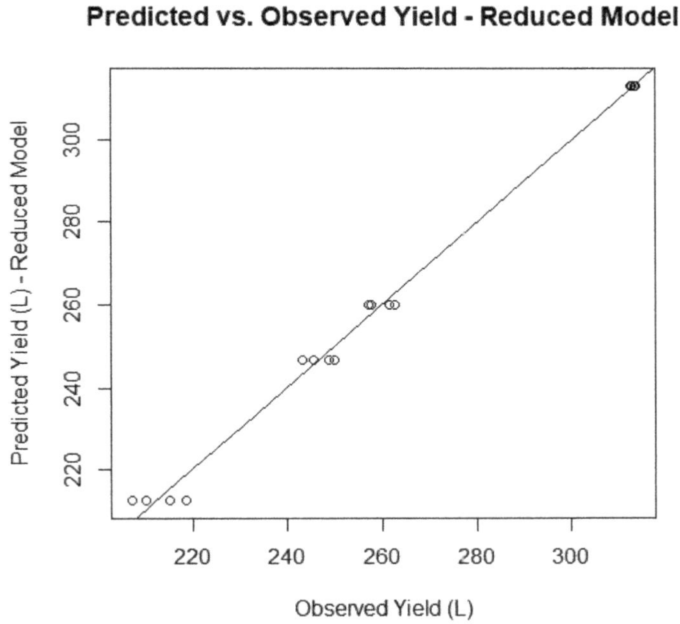

Fig. 12.2 Predicted versus observed yields

12.1 Reducing the Number of Runs—Fractional Factorials

The 2-level experimental design helps to reduce the numbers of runs required to assess the effects of factors. However, with more than four factors, even a 2-level experiment could be prohibitively expensive. For example, a 5-factor, 2-level experiment would have $2^5 = 32$ different runs. There is a method for doing only a fraction of the total number of runs in a 2^k experiment and still get the most relevant information. Such experiments are referred to as fractional factorial experiments (Box, et al., 1978). The idea is to select runs so that the information lost is about higher-order interactions (say 3-factor or higher orders of interaction) that are much less likely to matter enough to change any decisions.

Consider a modified version of the Yield experiment. Suppose in addition to the three original factors, temperature, pressure, and time, there were two others, say speed of the impeller in the reactor and the pH of the buffer material. A full 5-factor experiment would require $2^5 = 32$ runs, or combinations of the factor levels.

Before we introduce the way in which fractions are selected, we will consider a transformation, or coding, of factor levels. Recall that the factors all have a "low" level and a "high" level to be used in the experiment. A convenient way to code the factor levels to facilitate fraction selection is called Helmert coding (Box, et al. 1978). The levels are transformed onto the interval $(-1,+1)$. That is, in the case of the 2-level experiments, the low level becomes −1 and the high level becomes +1. If X represents the factor level in natural units, then the coding is achieved through:

$$\eta = \frac{X - m}{\frac{1}{2}R}$$

where:

$$m = \frac{X_L + X_H}{2}$$

and:

$$R = X_H - X_L$$

That is, m is the middle point between the low and high levels of factor X, and R is the range.

The multiple regression model can be fit to the coded levels, as long as you remember to code any other values of the factors if you want to get a prediction at some in-between set of conditions. To select a half-fraction, write down the total number of level combinations, then construct another set of columns for computing the coded levels. You choose a half-fraction, compute the highest order interaction in coded units by multiplying all the coded values. This new "column" will be all −1s and +1s. Furthermore, half the rows of the new column will be −1 and half will be plus one. To select a fraction, simply choose only those runs where the highest-order coded interaction is either +1 or −1.

Table 12.6 provides all $2^5 = 32$ runs for the 5-factor experiment, in both natural and Helmert-coded level values. Then the 5-factor interaction column is computed, with the rows that correspond to +1 being shaded.

The experimenter would choose either the 16 "−1" rows or "+1" rows. In either case, the experiment is a half-fraction of the full 2^5 experiment. Thus, each fraction is referred to as a 2^{5-1} fractional factorial experiment.

As mentioned earlier, something is "lost" when only a fraction of the total number of level combinations is performed. That loss is referred to as aliasing. That means some of the = potential effects cannot be distinguished from each other. In the case of fractional factorials, the aliasing is controlled. For a 2^{5-1} experiment, only interactions are aliased with each other. All the main effects are independent of any other effects. However, 2-factor interactions will be aliased with 3-factor interactions, but not with other 2-factor interactions. Inasmuch as the higher the order of interaction, the less influence it will have on the response, the tradeoff of information loss and reduction of experimental cost is well-justified generally.

Consider a half-fraction of the 5-factor Yield experiment. Table 12.7 provides the runs used for this experiment in terms of Helmert-coded factor values. Note that the column of the pH factor is identical to the 4-factor interaction Temp*Pres*Time*Speed. Hence the estimation of the pH effect and any

Table 12.6 $2^5 = 32$ Runs for the full 5-factor experiment

Natural Units						Coded Units					
Run	Temp	Pres	Time	Speed	pH	Temp	Pres	Time	Speed	pH	5-factor Interaction
1	40	10	30	60	7.2	-1	-1	-1	-1	1	-1
2	40	10	30	120	6.8	-1	-1	-1	1	-1	1
3	40	10	40	60	6.8	-1	-1	1	-1	-1	-1
4	40	10	40	120	7.2	-1	-1	1	1	1	1
5	40	14	30	60	6.8	-1	1	-1	-1	-1	1
6	40	14	30	120	7.2	-1	1	-1	1	1	-1
7	40	14	40	60	7.2	-1	1	1	-1	1	1
8	40	14	40	120	6.8	-1	1	1	1	-1	-1
9	60	10	30	60	6.8	1	-1	-1	-1	-1	-1
10	60	10	30	120	7.2	1	-1	-1	1	1	1
11	60	10	40	60	7.2	1	-1	1	-1	1	-1
12	60	10	40	120	6.8	1	-1	1	1	-1	1
13	60	14	30	60	7.2	1	1	-1	-1	1	1
14	60	14	30	120	6.8	1	1	-1	1	-1	-1
15	60	14	40	60	6.8	1	1	1	-1	-1	1
16	60	14	40	120	7.2	1	1	1	1	1	-1
17	40	10	30	60	7.2	-1	-1	-1	-1	1	1
18	40	10	30	120	6.8	-1	-1	-1	1	-1	-1
19	40	10	40	60	6.8	-1	-1	1	-1	-1	1
20	40	10	40	120	7.2	-1	-1	1	1	1	-1
21	40	14	30	60	6.8	-1	1	-1	-1	-1	-1
22	40	14	30	120	7.2	-1	1	-1	1	1	1
23	40	14	40	60	7.2	-1	1	1	-1	1	-1
24	40	14	40	120	6.8	-1	1	1	1	-1	1
25	60	10	30	60	6.8	1	-1	-1	-1	-1	1
26	60	10	30	120	7.2	1	-1	-1	1	1	-1
27	60	10	40	60	7.2	1	-1	1	-1	1	1
28	60	10	40	120	6.8	1	-1	1	1	-1	-1
29	60	14	30	60	7.2	1	1	-1	-1	1	-1
30	60	14	30	120	6.8	1	1	-1	1	-1	1
31	60	14	40	60	6.8	1	1	1	-1	-1	-1
32	60	14	40	120	7.2	1	1	1	1	1	1

Table 12.7 The 5-factor Half-fraction Experimental Design with the Temp*Pres*Time*Speed and the pH*Speed Interaction Columns Shown

Run	Temp	Pres	Time	Speed	pH	Temp*Pres*Time*Speed	pH*Speed
2	-1	-1	-1	-1	1	1	-1
3	-1	-1	-1	1	-1	-1	-1
5	-1	-1	1	-1	-1	-1	1
8	-1	-1	1	1	1	1	1
9	-1	1	-1	-1	-1	-1	1
12	-1	1	-1	1	1	1	1
14	-1	1	1	-1	1	1	-1
15	-1	1	1	1	-1	-1	-1
17	1	-1	-1	-1	-1	-1	1
20	1	-1	-1	1	1	1	1
22	1	-1	1	-1	1	1	-1
23	1	-1	1	1	-1	-1	-1
26	1	1	-1	-1	1	1	-1
27	1	1	-1	1	-1	-1	-1
29	1	1	1	-1	-1	-1	1
32	1	1	1	1	1	1	1

significance tests performed would be identical to those done for the Temp*Pres*Time*Speed interaction effect. That means pH is completely aliased with that 4-factor interaction. However, if the pH factor is significant, it is much more likely that this "main" effect is truly influencing the response, and not the 4-factor interaction. Conversely, the pH*Speed column is NOT identical to the pH column, which means that can be estimated independently of each other.

Table 12.8 provides the ANOVA for the 5-factor, half-fraction experiment. The letter "f" in front of each factor name indicates that for the ANOVA, the levels of each factor were treated as discrete.

The three main effects, Temp, Pres, and Time were all significant, whereas the main effects of Speed and pH were not. However, the interactions of Temp and pH, and Pres and Speed were.

Table 12.8 Yield 2^{5-1} experiment ANOVA

(Code 12.2)

Analysis of Variance Table

Response: Yield

	Df	Sum Sq	Mean Sq	F value	Pr(>F)
fTemp	1	51468	51468	6095.3180	< 2.2e-16 ***
fPres	1	28218	28218	3341.8111	< 2.2e-16 ***
fTime	1	13178	13178	1560.6925	< 2.2e-16 ***
fSpeed	1	24	24	2.8581	0.1082
fpH	1	0	0	0.0041	0.9498
fTemp:fPres	1	2616	2616	309.8083	8.633e-13 ***
fTemp:fTime	1	0	0	0.0002	0.9899
fPres:fTime	1	1	1	0.1319	0.7207
fTemp:fSpeed	1	8	8	0.8995	0.3555
fTemp:fpH	1	1057	1057	125.1748	1.545e-09 ***
fPres:fSpeed	1	609	609	72.1339	1.035e-07 ***
fPres:fpH	1	13	13	1.4818	0.2392
fSpeed:fpH	1	10	10	1.2380	0.2805
Residuals	18	152	8		

Signif. codes: 0 '***' 0.001 '**' 0.01 '*' 0.05 '.' 0.1 ' ' 1

Table 12.9 provides the multiple regression results, where each factor was treated as a continuously-valued regressor.

The same main effects and interactions that were significant in the ANOVA are significant in the regression.

Figure 12.3 shows the predicted Yield values plotted against the actual values, together with a regression line.

It appears that the predicted values are "close" to the actuals. The aliasing does not appear to severely affected the value of the experiment.

Other, smaller fractions are possible, especially when the number of factors is greater than 5. The smaller the fraction, the more intense the aliasing. There is an hierarchy of fractions and aliasing. They are referred to as "resolutions" and indicated with Roman numeral. Resolution III (ResIII) experiments have all interaction aliased with each other and with main effects, but all the main effects are independent of each other. They are useful if there are too many factors to experiment with, so an initial experiment can be performed to at least narrow the numbers of factors with which to perform other experiments. ResIV experiments have 2-factor interactions aliased with each other but not with main effects. ResIV experiments are more

Table 12.9 Regression results 2^{5-1} experiment

(Code 12.2)

Coefficients	Estimate	Std.Error	t value	Pr(>|t|)	
(Intercept)	99.43594	0.513685	193.574	2.00E-16	***
Temp	40.10469	0.513685	78.073	2.00E-16	***
Pres	29.69531	0.513685	57.808	2.00E-16	***
Time	20.29344	0.513685	39.506	2.00E-16	***
Speed	0.868438	0.513685	1.691	0.108	
pH	0.032812	0.513685	0.064	0.95	
Temp:Pres	9.041562	0.513685	17.601	8.63E-13	***
Temp:Time	-0.00656	0.513685	-0.013	0.99	
Pres:Time	0.186562	0.513685	0.363	0.721	
Temp:Speed	0.487187	0.513685	0.948	0.355	
Temp:pH	-5.74719	0.513685	-11.188	1.55E-09	***
Pres:Speed	4.362812	0.513685	8.493	1.03E-07	***
Pres:pH	-0.62531	0.513685	-1.217	0.239	
Speed:pH	0.571563	0.513685	1.113	0.28	

Signif. codes: 0 '***' 0.001 '**' 0.01 '*' 0.05 '.' 0.1 ' ' 1

Residual standard error: 2.906 on 18 degrees of freedom

Multiple R-squared: 0.9984, Adjusted R-squared: 0.9973

F-statistic: 885.5 on 13 and 18 DF, p-value: < 2.2e-16

(Code 12.2)

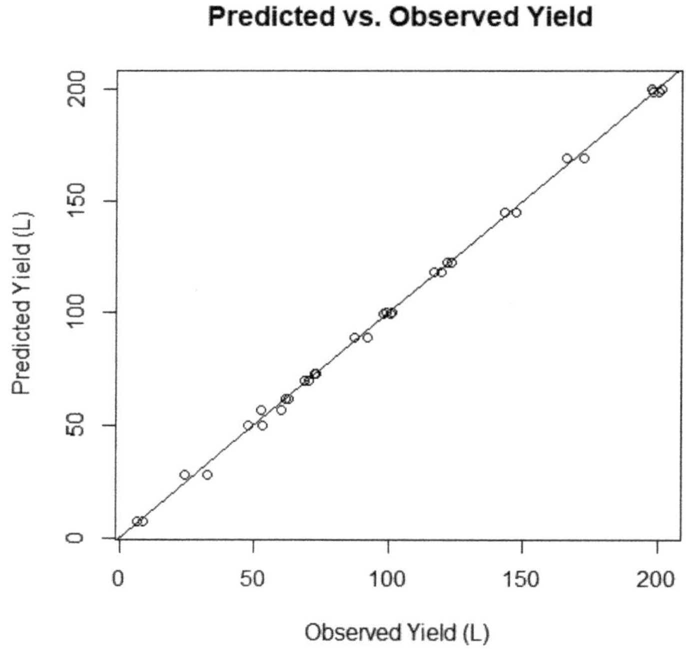

Fig. 12.3 Predicted versus actual yield—half-fraction experiment

useful if the experiment knows a priori that certain 2-factor interactions cannot actually exist, or at least are suspected to not be meaningful. ResV experiments, such as the 2^{5-1} experiment, have main effects independent of 2-factor interactions, and 2-factor interactions independent of each other.

Montgomery (2001) gives an excellent and detailed explanation of how to choose fractional designs to achieve particular aliasing structures.

12.2 Summary

Although ANOVA can be used with any number of factors each having different numbers of levels, it is sometimes useful to perform experiments, especially with factors whose values lie on a continuum, where each factor can only take two of the possible values, usually the smallest and largest for the factor. Such experiments are referred to as 2^k factorial experiments, where k = the number of factors/regressors to be included.

As the name implies, the number of combinations of factor levels is 2^k. In the case of $k = 3$, this is only $2^3 = 8$ combinations, or runs, required. If $k = 5$, then the number is much larger, namely $2^5 = 32$. It is possible to perform a fraction of the total number of possible runs, and still get sufficient information. Such experiments are called fractional factorials. The methodology for selecting fractions provides a means of controlling the information loss. Thus a more economical set of experiments can be performed and provide sufficient information to the decision maker/experimenter.

12.3 Points to Ponder

1. Is it justifiable to use a ResIII experiment when there are 11 factors, some of which may be important and some may not be?
2. How many experimental runs are required for a half-fraction of a 2^5 factorial experiment?
3. Convince yourself that all the two-way interactions in the half-fraction yield experiment are independent of each other.

References

Box, G. E. P., Hunter, W. G., & Hunter, J. S. (1978). *Statistics for experimenters*. Wiley.
Draper, N.R., & Smith, H. (1981). Applied Regression Analysis (2nd Ed.) John Wiley and Sons, New York.
Montgomery, D. C. (2001). *Design and analysis of experiments*. Wiley.

Chapter 13
Nonparametric Methods—Rank-Based Tests, Permutation Tests, and Resampling Methods

13.1 Why Nonparametrics?

Many conventional statistical methods rely on an assumption about the distributions from which the data are drawn. Notably, the t-test, the F-test (ANOVA), and even tests of hypotheses about a single standard deviation, all require that the data are drawn from normal distributions. Not only are the distributional assumptions potentially violated, but also in many cases, it is not possible to assess the degree to which those assumptions are violated.

There are techniques that are similar in goals to the more conventional methods but do not require the restrictive distributional assumptions. While they may seem very appealing, they are, in general, not as powerful when those assumptions are either not violated at all or are at least reasonably approximate. Nevertheless, the so-called nonparametric methods are very useful. There are two classes of nonparametric methods that will be focused on in this chapter:

1. Methods based on the rank transformation;
2. Resampling methods, such as bootstrap and permutation tests.

13.2 Rank-Based Methods

The basic idea of rank-transformation is to first order the data, regardless of any groupings, and number, or rank, them from smallest (1) to largest (N). Then replace the data with their respective ranks. Then form a statistic using the ranks instead of

---

Supplementary Information The online version contains supplementary material available at https://doi.org/10.1007/978-3-031-84479-9_13. The videos can be accessed individually by clicking the DOI link in the accompanying figure caption or by scanning this link with the SN More Media App.

S. A. Pardo, *Beyond the T-Test*, https://doi.org/10.1007/978-3-031-84479-9_13

the original values. Conover (1999) describes most of the common rank-based methods. Only two will be described here.

13.3 The Mann-Whitney-Wilcoxon Test: Analogous to the T-Test

A rank-based method analogous to a two-sample t-test is the Mann-Whitney-Wilcoxon (MWW) test. The procedure is simple:

1. Obtain two groups of data. The sample sizes are not necessarily identical. Say that for group 1, you obtain n values, and the group 2, m values. $N = n + m$.
2. Take all the N values and sort them from smallest to largest. The rank is the order in the sorted list.
3. Let $R(x_i)$ = the rank of value x_i (this is regardless of the group from which it came).
4. Just for group 1, compute:

$$U = \sum_{i=1}^{n} R(x_i)$$

5. The critical values for U will result in p-values that can be computed using the R function wilcox.test() (in base R). It allows the user to specify whether the test is lower tail, upper tail, or two-sided (default).

The degree to which the assumptions of the Student's t-test (or Welch's variant that does not require equal variances between the groups) will determine the amount of discrepancy between the conventional test and the rank-based test.

The Mann-Whitney-Wilcoxon test could be used to compare the two media in the OD experiment with bacteria cultures. Table 13.1 provides the output.

Table 13.1 The Mann-Whitney-Wilcoxon test—OD and media

```
> wilcox.test(OD ~ factor(MEDIUM))

        Wilcoxon rank sum test with continuity correction

data:  OD by factor(MEDIUM)
W = 151.5, p-value = 0.1096
alternative hypothesis: true location shift is not equal to 0

Warning message:
In wilcox.test.default(x = DATA[[1L]], y = DATA[[2L]], ...) :
  cannot compute exact p-value with ties
```

13.4 One-Way Nonparametric ANOVA: The Kruskal–Wallis Test

When there are more than two groups to compare, as in the case of the one-way layout, the MWW test is not adequate. Another rank-based test, called the Kruskal-Wallis test (Conover, 1999) can be used. The ranks are computed regardless of the group or treatment/condition under which the data were observed. So if x_{ij} is the jth observation in group i, then the ranks are $R(x_{ij})$. For each of k treatment groups, compute the sum of the ranks:

$$R_i = \sum_{j=1}^{n_i} R\left(x_{ij}\right)$$

The treatment groups are not required to have the same number of observations. Then compute the total sample size:

$$N = \sum_{i=1}^{k} n_i$$

and a variance estimate:

$$S^2 = \frac{1}{N-1}\left(\sum_{i,j} R\left(x_{ij}\right)^2 - \frac{N\left(N+1\right)^2}{4}\right)$$

The test statistic is:

$$\tau^2 = S^{-2}\left(\sum_{i=1}^{k} \frac{R_i^2}{n_i} - \frac{N\left(N+1\right)^2}{4}\right)$$

The test statistic approximately has, under the null hypothesis of no difference (on the average) between groups, a Chi-squared distribution with $k - 1$ degrees of freedom. If it is larger than the $100(1 - \alpha)$ percentile of the Chi-squared distribution having $k - 1$ degrees of freedom, then reject the null hypothesis of no difference.

In a modified experiment with optical density (OD), suppose that a third PREP was included, in between liquid and solid. Call this in-between level "slush."

A Kruskal-Wallis test could be performed with OD as a function of PREP only.

Table 13.2 provides the output for the Kruskal-Wallis test, using only the factor PREP.

Table 13.2 Kruskal-Wallis test—OD ~ PREP

```
> kruskal.test(OD ~ PREP)

        Kruskal-Wallis rank sum test

data:  OD by PREP
Kruskal-Wallis chi-squared = 21.838, df = 2, p-value = 1.811e-05
```

ANOVA on Ranks

The Mann-Whitney-Wilcoxon and Kruskal-Wallis tests cover the two-group t-tests and the one-way layout with more than two groups. But what if there are multiple factors, as in the case of the OD experiment? If there is concern that the assumptions of either normally distributed errors (noise) or equal variances for each grouping of data, it is possible to use the rank transformation on the data and then perform the multi-factor ANOVA (Conover, 1999) The ranks are used instead of the raw data. Such a transformation will reduce the influence of a small number if values that are very far from the group average ("outliers") and hopefully give a better indication of what differences may truly exist.

An interaction plot can be constructed for the modified OD experiment. Figure 13.1 shows the plot.

This plot is in the original OD units. Another interaction plot with the ranks of OD is shown in Fig. 13.2.

Table 13.3 provides the ANOVA output using the Ranks in place of the raw OD data.

Compare this to the ANOVA for the modified experiment on the raw OD Data, given in Table 13.4.

The interaction effect was significant using the raw data but not with the ranks. This discrepancy shows how care must be taken before using ranks instead of the raw data. There should be some graphical representations of the data used before it is decided that the rank transformation is appropriate.

(Code 13.1)

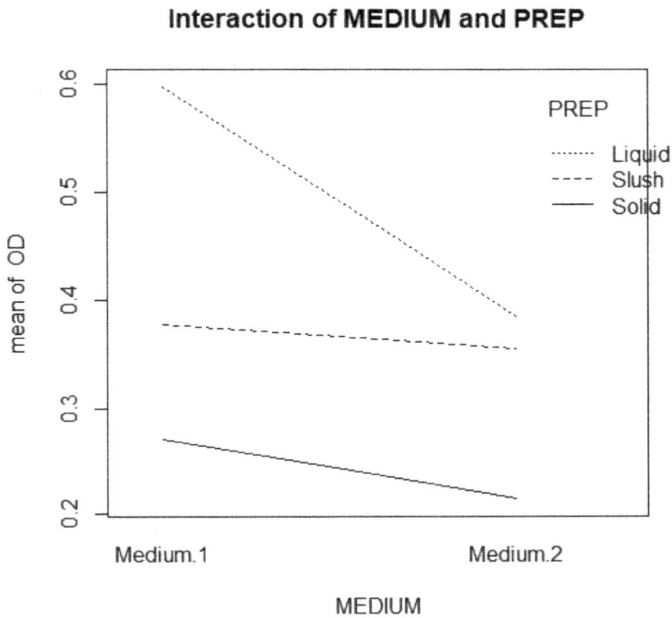

Fig. 13.1 Interaction plot—modified OD experiment

(Code 13.1)

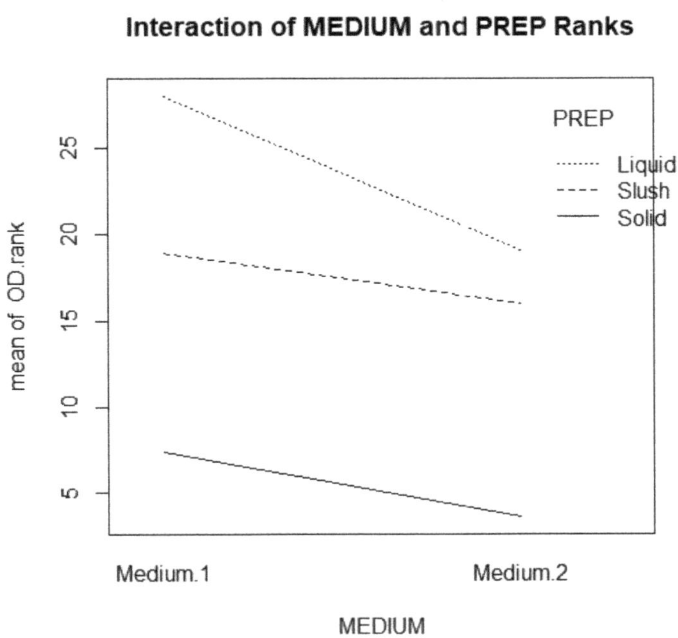

Fig. 13.2 Interaction Plot—Modified OD Experiment—Ranks

Table 13.3 ANOVA for OD Ranks

(Code 13.1)

Analysis of Variance Table

Response: OD.rank

	Df	Sum Sq	Mean Sq	F value	Pr(>F)
factor(MEDIUM)	1	202.80	202.80	16.3439	0.0004729 ***
factor(PREP)	2	1686.05	843.02	67.9402	1.309e-10 ***
factor(MEDIUM):factor(PREP)	2	52.35	26.17	2.1095	0.1432342
Residuals	24	297.80	12.41		

Signif. codes: 0 '***' 0.001 '**' 0.01 '*' 0.05 '.' 0.1 ' ' 1

Table 13.4 ANOVA for OD—modified experiment

(Code 13.1)

Analysis of Variance Table

Response: OD

	Df	Sum Sq	Mean Sq	F value	Pr(>F)
factor(MEDIUM)	1	0.070083	0.070083	36.002	3.406e-06 ***
factor(PREP)	2	0.312500	0.156250	80.265	2.343e-11 ***
factor(MEDIUM):factor(PREP)	2	0.051327	0.025663	13.183	0.000137 ***
Residuals	24	0.046720	0.001947		

Signif. codes: 0 '***' 0.001 '**' 0.01 '*' 0.05 '.' 0.1 ' ' 1

13.5 Resampling Methods: Bootstrap

The bootstrap was introduced in Chap. 4 and 8 for computing confidence intervals on differences of proportions or odds ratios, neither of which have any simple closed-form expressions for confidence intervals. The bootstrap approach can be used to test hypotheses and compute confidence intervals when parametric assumptions are either violated or their status is unknown.

13.6 Resampling Methods: Permutation Test

Permutation tests (Good, 1994) are useful, especially when the test statistic's sampling distribution is either unknown or intractable. The procedure is as follows:

1. Compute the test statistic with the "unpermuted" data;
2. For N repetitions:
   (a) Permute the data by randomly sampling (with replacement) and randomly assigning each value to a treatment group;
   (b) Use the permuted data to compute the test statistic, and store its value;
3. Find the tail that the original test statistic falls in the distribution of the permuted data statistics;

The tail probability gives the *p*-value. The permuted data distribution is the "null" distribution, inasmuch as the assignment of data to treatment groups was random and therefore meaningless. Figure 13.3 shows a histogram of permuted data from the example in Chap. 8, used to compute the null distribution for the difference in proportions, PA – PB. The vertical line shows where the observed difference of PA – PB = 0.09 lies in this distribution. Its *p*-value is approximately 0.015, which is significant at the 0.05 level.

(Code 8.2)

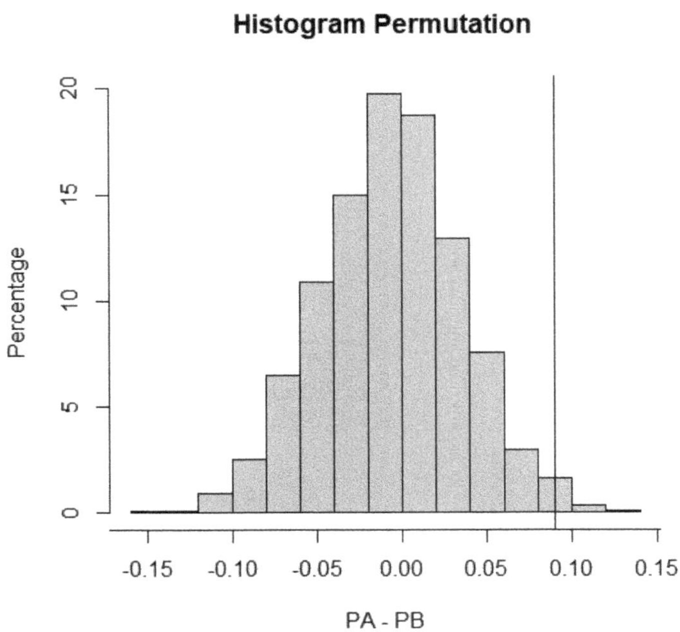

Fig. 13.3 Permutation "Null" distribution for differences in proportions of successes

Permutation tests can be used for any kind of parameter, either from a discrete or continuous distribution.

13.7 Contingency Tables

Two discrete variables may or may not be related. For example, an organization might be interested in knowing whether gender is related to whether a member is in a managerial position. Typically an hypothesis involves wither the distribution of one variable is independent of the value of the other variable. One might hypothesize that the frequency of a value of one variable is uniformly distributed across all the values of the other. Data for the joint distribution of two discrete variables can be arrayed in what is called a contingency table (Conover, 1999). Suppose there are $n = 100$ employees in company X, with 40 men and 60 women. Furthermore, suppose there are 30 managerial positions (we are presuming here that a position is either managerial or not). So there are 100 positions; 30 managerial and 70 not. Thus, if there is no bias toward a particular gender being in a managerial position, we would expect that about 40% of the managerial positions (0.4*30 = 12) are occupied by men and 60% of the managerial positions (0.6*30 = 18) are occupied by women. That would leave 40–12 = 28 men in non-managerial positions and 60–18 = 42 women in non-managerial positions. Table 13.5 provides the expected numbers under the hypothesis of no bias.

Table 13.6 provides the actual joint distribution.

To test the hypothesis, we compute the following statistic:

$$\hat{\chi}^2 = \sum_{i=1}^{k}\sum_{j=1}^{l}\frac{\left(O_{ij} - E_{ij}\right)^2}{E_{ij}}$$

Table 13.5 Expected joint distribution gender × position type

	H0: Rows and Columns Independent Expected Joint Distribution under H0: Gender x Position Type		
	Managerial	Non-Managerial	Row Total
Men	12	28	40
Women	18	42	60
Col Total	30	70	100

Table 13.6 Actual joint distribution gender × position type

	H0: Rows and Columns Independent Actual Joint Distribution: Gender x Position Type		
	Managerial	Non-Managerial	Row Total
Men	4	36	40
Women	26	34	60
Col Total	30	70	100

Table 13.7 Calculations for the Chi-squared Statistic

	H0: Rows and Columns Independent ((Oberved-Expected)**2)/Expected: Gender x Position Type		
	Managerial	Non-Managerial	Row Total
Men	16.0	2.3	18.3
Women	3.6	1.5	5.1
Col Total	19.6	3.8	**23.4**
			χ2 Statistic

where O represents the number of observed results in cell i, j, and E the number of expected results if the null hypothesis H0 is true. In this case, $k = 2$ and $l = 2$. That is, there are $k = 2$ genders and $l = 2$ position types.

As the symbol implies, this statistic has approximately a Chi-squared distribution if the null hypothesis (H0: rows and columns are uncorrelated) is true, with degrees of freedom $(k - 1)*(l - 1) = 1$ in this case. Table 13.7 provides the computations of the $\frac{(O_{ij} - E_{ij})^2}{E_{ij}}$ and the resultant sum.

We can compute the p-value, namely the probability that a Chi-squared random variable with 1 degree of freedom exceeds the computed value of 23.4. That p-value is approximately $1.39*10^{-6} \approx 0$. So we reject H0 and conclude that there is a relationship between gender and position type.

While the Chi-squared statistic is "approximate," there is an "exact" test for contingency tables, called Fisher's Exact Test (Conover, 1999) The output for this table (Table 13.6) is given in Table 13.8. The only issue with Fisher's Exact Test is computational speed. It is generally not practical if k or l are greater than 2. That is, Fisher's Exact Test is almost exclusively good for only 2 × 2 tables. However, many algorithms use a resampling approach to computing p-values for Fisher's Exact Test, which make them faster than they would be otherwise.

Although the p-value for the Chi-squared statistic is lower than that of the Fisher's Exact test, both were well lower than 0.05, indicating that H0 should be rejected.

The power for a given sample size in creating a continency table can be computed using something called a non-central $\chi 2$ distribution. This is a generalization of the $\chi 2$ distribution, with an extra parameter called non-centrality, equal to:

$$ncp = \omega^2 n$$

The symbol ω represents the population "correlation" between the two and column variables, which can range from 0, independence, to 1, perfect correlation (Yule, 1912). A correlation of 0 implies H0 is true.

Figure 13.4 shows a power curve for $n = 100$ with $k = l = 2$.

Table 13.8 Fisher's exact test for gender × position type table

(Code 13.2)

```
> data.mat
          Position
Gender  Mgr Non-Mgr
  Men      4    36
  Women   26    34
> test.stat <- fisher.test(x=data.mat)
> test.stat

        Fisher's Exact Test for Count Data

data:  data.mat
p-value = 0.0003272
alternative hypothesis: true odds ratio is not equal to 1
95 percent confidence interval:
 0.03395627 0.48924522
sample estimates:
odds ratio
 0.1479556
```

(Code 13.3)

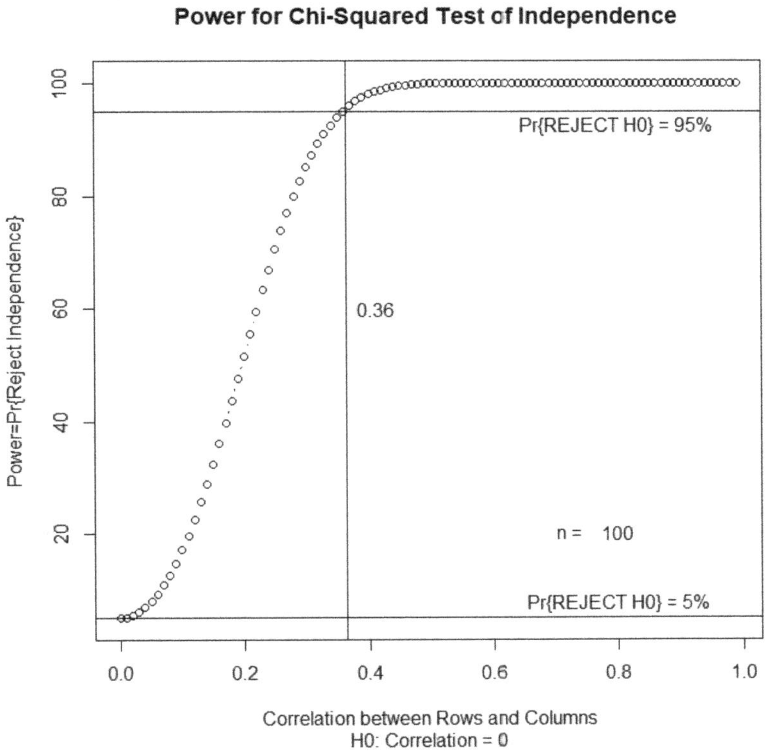

Fig. 13.4 Power Curve for $k \times l$ contingency table with $n = 100$

13.8 Summary

Nonparametric methods (in particular, rank-based tests) are useful when assumptions required for parametric tests are violated, and methods such as bootstrapping and permutation tests are useful when the sampling distributions of test statistics are either unknown or intractable. In the case of rank tests, it may be reasonable to replace the original data with their ranks and perform more conventional tests on the ranks. One of the uses of the rank transformation is that it reduces the degree to which highly disparate response values affect the result. This is particularly true of ANOVA, which relies on having equal variances for all treatment groups. Bootstrap and permutation tests are very useful for situations where assumptions are violated, or where the test statistic of interest has an unknown or mathematically intractable sampling distribution.

Sometimes, it is of interest to know if two discrete variables are related to each other. A contingency Table analysis with a Chi-squared test statistic can be used to decide whether the two variables are independent (uncorrelated).

13.9 Points to Ponder

1. Under what conditions would you think a rank-based method would be appropriate?
2. Can you envision using a permutation test methodology for ANOVA? How would you proceed? As an example, consider a one-way design.

References

Conover, W. J. (1999). *Practical nonparametric statistics* (3rd ed.). Wiley.

Good, P. (1994). *Permutation tests*. Springer.

Yule, G. U. (1912). On the methods of measuring association between two attributes. *Journal of the Royal Statistical Society., 75*(6), 579–652.

Chapter 14
Nonlinear and Logistic Regression

14.1 Nonlinearity in Models

The regressions described so far have all been linear in parameters. That is, the model form has always been:

$$y = \beta_0 + \beta_1 X_1 + \beta_2 X_2 + \cdots + \beta_k X_k + \varepsilon = \beta_0 + \sum_{i=1}^{k} \beta_i X_i + \varepsilon$$

Even if any of the regressors have a "nonlinear" form, e.g.:

$$X_i = W^3$$
$$X_i = 2.5\sin(W)$$
$$X_i = \log(W)$$

As long as W is a known quantity, all of these regressors could still be part of a linear model, inasmuch as the coefficients are simply unknown multipliers to be estimated.

However, suppose that the model has a term of the form:

$$\sin\left(\beta_i X_i^{\gamma_i}\right)$$

where β_i and γ_i are unknown parameters to be estimated, then the model is no longer linear. There are techniques to estimate the parameters of such models. We will refer to these methods as nonlinear regression (Seber & Wild, 1989).

Supplementary Information The online version contains supplementary material available at https://doi.org/10.1007/978-3-031-84479-9_14. The videos can be accessed individually by clicking the DOI link in the accompanying figure caption or by scanning this link with the SN More Media App.

S. A. Pardo, *Beyond the T-Test*, https://doi.org/10.1007/978-3-031-84479-9_14

There are some models that can be transformed to allow for the use of linear least squares. There are some, however, where it is not possible. We will discuss both situations.

14.2 Linearizable Models

In some cases, although the model is nonlinear, a transformation can allow linear least squares to estimate the parameters.

Consider a very simple example. Suppose that a sterilization method will reduce bacteria colonies at an exponential rate. If Y_t is the number of colonies remaining after t time units of exposure, and if y_0 is the initial colony count, then:

$$Y_t = y_0 e^{-\lambda t+\varepsilon}$$

The unknown parameter, λ, cannot be estimated via linear least squares in this form. However, a simple transformation can be made to "linearize" the model:

$$\ln Y_t = \ln y_0 - \lambda t + \varepsilon$$

In this linearized model, there is one regressor, t. There are two parameters to estimate:

$$\beta_0 = \ln y_0$$

$$\beta_1 = -\lambda$$

The response variable is the natural log of the number of colonies at time t. If the initial count is known, then β_0 does not require estimation. The response can be "adjusted" for the initial count:

$$Z = \ln Y_t - \ln y_0$$

Now least squares can be used to estimate the one unknown parameter. Fit the model:

$$Z = \beta_1 t + \varepsilon$$

Once the parameter is estimated, you can obtain predicted values:

$$\hat{Y}_t = y_0 e^{\hat{\beta}_1 t}$$

A generalization of this process was actually demonstrated in Chap. 11, with the computation of the cumulative hazard function:

$$-\ln \hat{R}(t) = \hat{H}(t)$$

and then modeling it as a polynomial in the time variable, t:

$$H(t) = \beta_0 + \beta_1 t + \beta_2 t^2 - \varepsilon$$

The values of $\hat{R}(t)$ can be obtained using the Kaplan–Meier statistics on time-to-occurrence data, as described in Chap. 11.

14.3 Generalized Linear Models—Logistic Regression

Along with linearilizable models, there is a more general notion of the generalized linear model (McCullagh & Nelder, 1989). The idea is that a nonlinear model may have two associated components:

1. The link function, which is the transformation to a linear form;
2. The distribution of errors, or noise, which may not be normal.

While there are many such models, the discussion here will be limited to one, namely logistic regression (Hosmer & Lemeshow, 1989). In a logistic regression, the response is binary. There is a generalization of this, called ordinal logistic regression, where the response is on an ordinal scale, such as a Likert scale (Likert, 1932), but we will not deal with that here. The model is in terms of the probability of a "success" as a function of the regressors. The form of the nonlinear model is:

$$\Pr\{Y = 1 | x_0, x_1, \cdots, x_k\} = \frac{\exp\left(\sum_{i=0}^{k} \beta_i x_i\right)}{1 + \exp\left(\sum_{i=0}^{k} \beta_i x_i\right)} = p_1$$

It turns out that:

$$p_1 = \frac{\exp\left(\sum_{i=0}^{k} \beta_i x_i\right)}{1 + \exp\left(\sum_{i=0}^{k} \beta_i x_i\right)} = \frac{1}{1 + \exp\left(-\sum_{i=0}^{k} \beta_i x_i\right)}$$

The inverse of p1 is conveniently:

The regressor x_0 is a "dummy" variable, set to 1 for all cases. Thus the parameter β_0 is just a constant "intercept."

The link function that linearizes the model is:

$$f(x_0, x_1, x_2, \ldots, x_k) = \ln\left(\frac{\Pr\{Y = 1 | x_0, x_1, \cdots, x_k\}}{1 - \Pr\{Y = 1 | x_0, x_1, \cdots, x_k\}}\right) = \ln\left(\frac{p_1}{(1 - p_1)}\right) = \sum_{i=0}^{k} \beta_i x_i$$

To estimate the values of the β_i, maximize the logarithm of the likelihood function:

$$L(\beta_0, \beta_1, \ldots, \beta_k) = \sum_{j=1}^{n} y_j \ln(p_1) + (1 - y_j)\ln(1 - p_1)$$

Recall that p_1 is a function of all the regressors, and

$$y_j = \begin{cases} 1\,\text{if success} \\ 0\,\text{if not success} \end{cases}$$

Since the response data are binary, the error distribution for the logistic model is binomial.

The optimization is performed via an iterative numerical method. It requires "convergence," which means that at some point the iterations do not change the resulting values of the parameter estimates by more than some small amount. If convergence is not achieved for some particular set of data, it means that the estimates are not necessarily reliable.

The logistic model is commonly used to predict the probability of success for different values of the regressors. It can also be used as a sort of ANOVA for discrete factors with a binary response.

Consider an example with ten continuously-valued regressors and a binary response. The analysis results are given in Table 14.1.

Consider an experiment with two industrial robots, designed to pick products from a production line and place them in the appropriate bin. An experiment was performed to determine how successful the two robots were at discriminating between the two products and placing them in the correct bin.

Table 14.2 provides the results.

Figure 14.1 shows an interaction plot between the robot designs and the products.

In both analyses, the z-value or z-score for the parameter estimate was computed as the estimate divided by its standard error. A significant z-score indicates that the parameter is not 0.

Table 14.1 Logistic regression with ten continuously-valued regressors

(Code 14.1)

Logistic Regression with Continuous Regressors

Term	Estimate	Std.Error	z-value	Pr(>|z|)	Code
(Intercept)	5.951372	2.580635	2.306	0.0211	*
X.01	-0.03366	0.058756	-0.573	0.56675	
X.02	0.677804	0.158421	4.278	1.88E-05	***
X.03	-0.30626	0.145995	-2.098	0.03593	*
X.04	-0.01975	0.004824	-4.094	4.25E-05	***
X.05	0.121358	0.099517	1.219	0.22267	
X.06	0.051108	0.048472	1.054	0.29171	
X.07	0.223852	0.10266	2.181	0.02922	*
X.08	0.019647	0.069061	0.284	0.77604	
X.09	-12.8093	4.200661	-3.049	0.00229	**
X.10	0.130501	0.079524	1.641	0.10079	

---

	p-value:	0	0.001	0.01	0.05	0.1
Signif.	code:	***	**	*	.	

Table 14.2 Logistic regression with discrete factors

(Code 14.2)

Term	Estimate	Std.Error	z-value	Pr(>|z|)	Code
(Intercept)	1.1527	0.4683	2.461	0.0138	*
factor(Robot)2	2.0254	1.1227	1.804	0.0712	.
factor(Product)2	-0.7472	0.6213	-1.203	0.2291	
factor(Robot)2:factor(Product)2	-0.7726	1.3133	-0.588	0.5563	

	p-value:	0	0.001	0.01	0.05	0.1
Signif.	code:	***	**	*	.	

(Code 14.2)

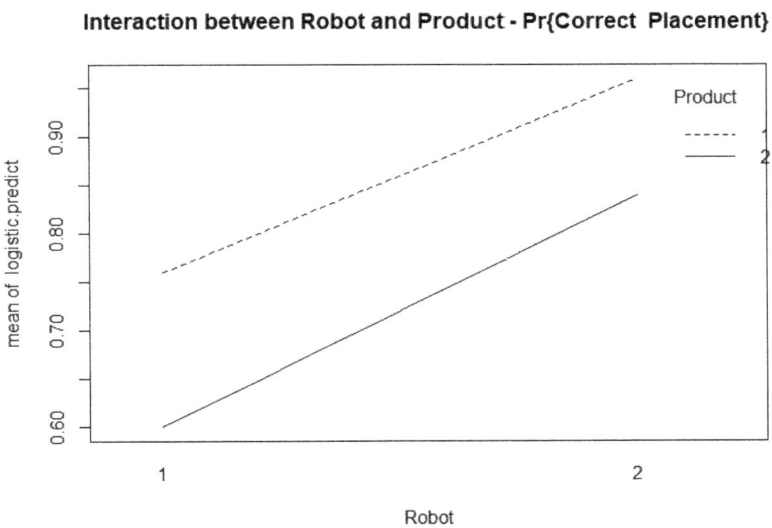

Fig. 14.1 Interaction plot for logistic regression with discrete factors

14.4 Unlinearizable Models

Sometimes a model is suggested that cannot be linearized. That is, there is no link function that is a linear function of parameters. In such cases, numerical iterative methods can be used to estimate parameters.

Example data, (X,Y) are plotted in Fig. 14.2. The form of model at might be appropriate could be:

Fig. 14.2 Nonlinear data **(Code 14.3)**

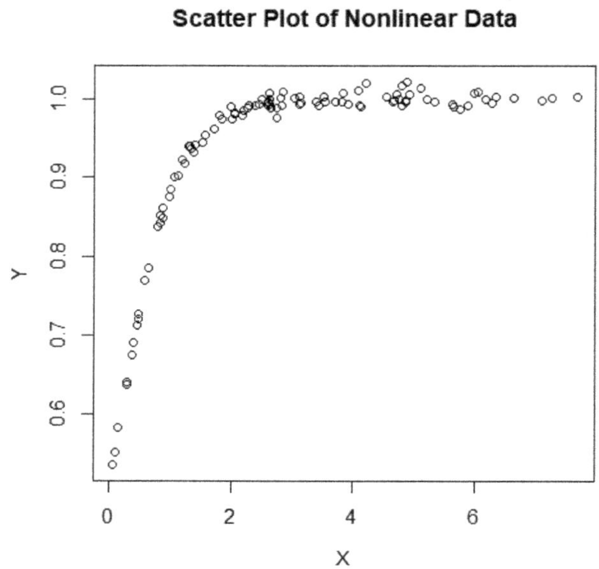

Fig. 14.3 Nonlinear data and fit values **(Code 14.3)**

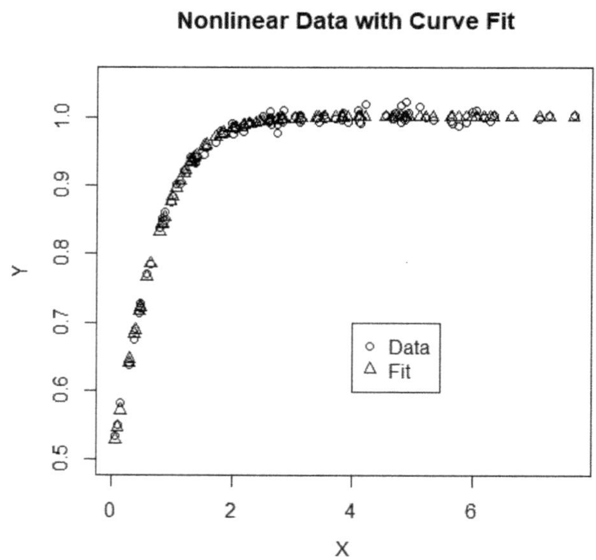

$$Y = \frac{\exp(\beta_0 X)}{1 + \exp(\beta_1 X)} + \varepsilon$$

Figure 14.3 shows the data together with predicted (fit) values. Table 14.3 provides the model with parameter estimates.

Table 14.3 Nonlinear regression model and parameter estimates

(Code 14.3)

Nonlinear Regression Model
Formula: Y ~ exp(B0 * X)/(1 + exp(B1 * X))

Parameters:

	Estimate	Std.Error	t.value	Pr(>|t|)
B0	2.02178	0.01464	138.1	<2e-16
B1	2.02164	0.01468	137.7	<2e-16

Signif. codes: 0 '***' 0.001 '**' 0.01 '*' 0.05 '.' 0.1 ' ' 1

Residual standard error: 0.007145 on 98 degrees of freedom

Number of iterations to convergence: 8
Achieved convergence tolerance: 5.772e-06

It took eight iterations of the nonlinear algorithm to converge. The convergence tolerance of $5.772*10^{-6}$ means that once the iterated estimate did not change by more that this tolerance, the algorithm stopped.

14.5 Summary

Linear regression works for models that are linear in their parameters. Even if the regressors themselves are not linear functions of other known variables, as long as the model parameters are multiplicative coefficients, linear regression applies. However, there are cases where the model parameters are not simply multipliers, but may be such that the model cannot be expressed as a linear combination of regressors, e.g.:

$$\beta_0 x_0 + \beta_1 x_1 + \ldots + \beta_k x_k$$

There are some special cases where the model is not expressly linear in parameters, but through a transformation, can be linearized so that the parameters can be estimated by linear least squares. There are another set of cases where, although the model cannot be simply linearized, a transformation, called a link function, creates a linear function that makes estimation easier. One of the most common such situations is that of logistic regression, where the response is binary. Finally, there are situations where no such link function exists. In those cases, numerical methods can be applied to estimate parameters.

14.6 Points to Ponder

1. What makes a model nonlinear?
2. How are parameters estimated for nonlinear models?
3. How can a binary variable be used to test whether discrete factors are significant?

References

Hosmer, D. W., & Lemeshow, S. (1989). *Applied logistic regression*. Wiley.

Likert, R. (1932). A technique for the measurement of attitudes. *Archives of Psychology, 140*, 1–55.

McCullagh, P., & Nelder, J. A. (1989). *Generalized linear models* (2nd ed.). Chapman and Hall/CRC.

Seber, G. A. F., & Wild, C. J. (1989). *Nonlinear regression*. Wiley.

Chapter 15
Model Building

There are times when there are many possible regressors or explanatory variables which may or may not actually affect the response. This situation can occur in the case of designed experiments, or in "observational" studies, where many variables on each experimental unit, but without controlling their levels.

15.1 Linear Models

In the cases of linear models or generalized linear models, stepwise regression (Draper & Smith, 1981) is a class of methods that can be used to sift through many regressors to determine the "final" group to be used for a predictive model. The idea is to either start with one regressor and slowly add and remove regressors based on some measure of goodness, such as Cp, until the improvement in the measure of goodness is at or below some predetermined value. This is called forward stepwise regression. Backward stepwise regression starts with all possible regressors and removes or adds back regressors based on the measure of model goodness, and terminates when the measure of goodness reaches some prespecified level.

Supplementary Information The online version contains supplementary material available at https://doi.org/10.1007/978-3-031-84479-9_15. The videos can be accessed individually by clicking the DOI link in the accompanying figure caption or by scanning this link with the SN More Media App.

S. A. Pardo, *Beyond the T-Test*, https://doi.org/10.1007/978-3-031-84479-9_15

15.2 Stepwise Regression—Continuous Response

The stepwise algorithms proceed until there is no longer a change in the measure of goodness larger than the tolerance, or threshold, set before the analysis. A common measure is called Akaike Information Criterion, or AIC. For each model in the stepwise process, AIC is:

$$\mathrm{AIC}_p = \frac{\mathrm{SSE}_p}{\hat{\sigma}^2} + \left(\ln\left(2\pi\right) + + \ln\left(\hat{\sigma}^2\right)\right)N + 2p$$

p = number of parameters in the model
N = sample size
$\hat{\sigma}^2$ = root mean squared error term from the model (an estimate of model error variance)
SSE_p = the sum of squared residuals from the model with p parameters.

The reason that AIC is used as opposed to adjusted R^2 is that as the number of parameters increases, adjusted R^2 will get closer and closer to 1, regardless of how good the model is as a predictive equation. For AIC, the lower it is, the "better" the model. Its cousin, C_p, has an ideal value equal to p, the number of parameters:

$$C_p = \frac{\mathrm{SSE}_p}{\hat{\sigma}^2} - N + 2p$$

As mentioned above, SSE_p is the sum of squared residuals with a model having p parameters. So:

$$\hat{\sigma}^2 = \frac{\mathrm{SSE}_p}{N - p}$$

Consider the example from Chap. 14 with ten continuous regressors, only this time with a continuous response.

Table 15.1 provides the results from a "usual" multiple regression.

There were five main effects that were significant (besides the intercept): X.01, X.03, X.06, X.07, X.08. Only one of the two-way interactions was significant: X.03:X.07.

A backward stepwise procedure was performed. The original (full) model with 26 parameters had an AIC of 99.55. After 22 "steps," or iterations, the final model was:

Response ~ X.01 + X.03 + X.06 + X.07 + X.08 + X.09 + X.10 + X.01:X.08 + X.0 3:X.07 + X.07:X.10

with an AIC of 69.05, and 11 parameters (including the intercept). Table 15.2 provides the parameter estimates for the final model terms.

Figure 15.1 shows the fitted (predicted) values of response plotted against the observed response values. The line is a simple linear regression of predicted ~ observed.

Table 15.1 Multiple regression with ten regressors and all two-way interactions

(Code 15.1)

```
Call:
lm(formula = Response ~ X.01 + X.02 + X.03 + X.04 + X.05 + X.06 +
  X.07 + X.08 + X.09 + X.10 + X.01:X.03 + X.01:X.06 + X.01:X.07 +
  X.01:X.08 + X.01:X.09 + X.01:X.10 + X.03:X.06 + X.03:X.07 +
  X.03:X.08 + X.03:X.09 + X.03:X.10 + X.06:X.07 + X.06:X.08 +
  X.06:X.09 + X.06:X.10 + X.07:X.08 + X.07:X.09 + X.07:X.10 +
  X.08:X.09 + X.08:X.10 + X.09:X.10)

Residuals:
  Min     1Q  Median    3Q    Max
-2.4587 -0.6820 -0.0538 0.6550 3.4495

Coefficients:
```

	Estimate	Std.Error	t.value	Pr(>|t|)	Code
(Intercept)	-1.19E+01	3.08E+00	-3.881	0.000149	***
X.01	-1.10E+00	1.41E-01	-7.813	5.77E-13	***
X.02	4.16E-02	4.45E-02	0.935	0.35125	
X.03	-2.70E+00	3.30E-01	-8.173	6.98E-14	***
X.04	-8.85E-04	1.42E-03	-0.622	0.534611	
X.05	2.95E-02	3.02E-02	0.977	0.33014	
X.06	5.03E+00	8.73E-02	57.632	< 2e-16	***
X.07	-2.01E+00	2.40E-01	-8.389	1.93E-14	***
X.08	-3.13E+00	1.21E-01	-25.932	< 2e-16	***
X.09	4.37E+00	6.24E+00	0.701	0.484349	
X.10	5.86E-02	1.63E-01	0.36	0.719177	
X.01:X.03	3.76E-03	9.37E-03	0.401	0.689194	
X.01:X.06	1.49E-03	3.23E-03	0.46	0.646255	
X.01:X.07	-6.68E-03	6.20E-03	-1.076	0.283388	
X.01:X.08	-4.97E-03	4.05E-03	-1.228	0.221217	
X.01:X.09	-3.30E-01	2.49E-01	-1.327	0.186444	
X.01:X.10	-1.98E-03	5.42E-03	-0.366	0.715099	
X.03:X.06	-1.51E-04	7.51E-03	-0.02	0.984043	
X.03:X.07	-4.67E-02	2.00E-02	-2.336	0.020654	*
X.03:X.08	2.70E-03	1.04E-02	0.261	0.794724	
X.03:X.09	1.53E-02	5.47E-01	0.028	0.97776	
X.03:X.10	-5.28E-03	1.34E-02	-0.395	0.693412	
X.06:X.07	3.71E-03	6.07E-03	0.611	0.541772	
X.06:X.08	7.39E-04	3.89E-03	0.19	0.84952	
X.06:X.09	-3.69E-02	1.88E-01	-0.196	0.844933	
X.06:X.10	-6.70E-04	4.21E-03	-0.159	0.87393	
X.07:X.08	-4.20E-03	6.14E-03	-0.685	0.494543	
X.07:X.09	-5.66E-01	4.88E-01	-1.16	0.247837	
X.07:X.10	1.45E-02	1.01E-02	1.431	0.154153	
X.08:X.09	-2.14E-01	2.57E-01	-0.833	0.406021	
X.08:X.10	-3.33E-03	6.07E-03	-0.549	0.583639	
X.09:X.10	1.84E-01	3.22E-01	0.57	0.569316	

```
---
Signif. codes: 0 '***' 0.001 '**' 0.01 '*' 0.05 '.' 0.1 ' ' 1

Residual standard error: 1.192 on 168 degrees of freedom
Multiple R-squared: 0.9991,  Adjusted R-squared: 0.9989
F-statistic: 6010 on 31 and 163 DF, p-value: < 2.2e-16
```

Table 15.2 Final terms and parameter estimates

(Code 15.2)

Term	Estimate
(Intercept)	-11.8456
X.01	-1.17118
X.03	-2.70447
X.06	5.008715
X.07	-2.21455
X.08	-3.19577
X.09	5.283319
X.10	0.094627
X.01:X.08	-0.00569
X.03:X.07	-0.04923
X.07:X.10	0.014431

(Code 15.2)

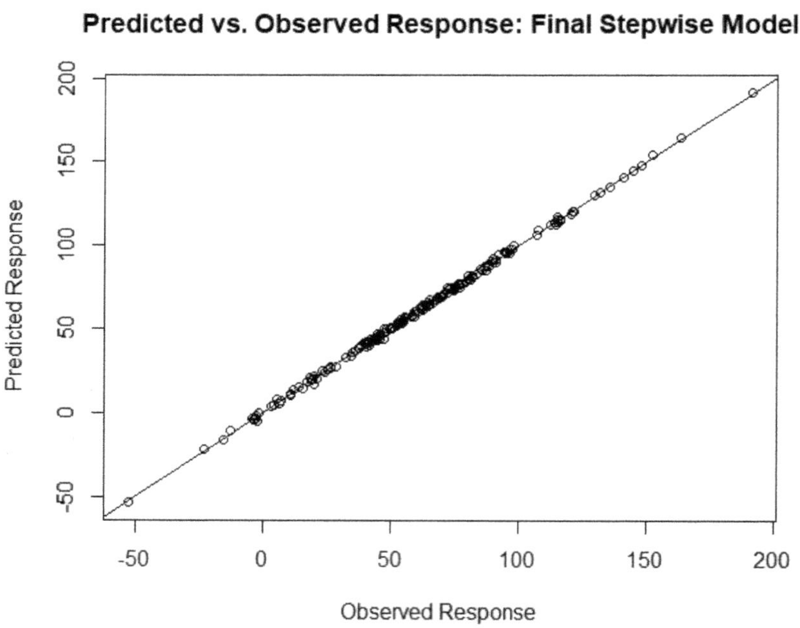

Fig. 15.1 Predicted response from final stepwise model vs. observed response

15.3 Stepwise Regression—Binary Response

Logistic regression can also be used in a stepwise procedure. The same ten regressors and their two-way interactions were fit with binary data. Table 15.3 provides the results.

There were 12 significant terms, at least at the 0.10 level. The full model had an AIC of 113.69. After 15 steps, the final model was:

Table 15.3 Logistic regression with ten regressors and all two-way interactions

(Code 15.2)

```
Call:
glm(formula = Response.B ~ X.01 + X.02 + X.03 + X.04 + X.05 +
  X.06 + X.07 + X.08 + X.09 + X.10 + X.01:X.03 + X.01:X.06 +
  X.01:X.07 + X.01:X.08 + X.01:X.09 + X.01:X.10 + X.03:X.06 +
  X.03:X.07 + X.03:X.08 + X.03:X.09 + X.03:X.10 + X.06:X.07 +
  X.06:X.08 + X.06:X.09 + X.06:X.10 + X.07:X.08 + X.07:X.09 +
  X.07:X.10 + X.08:X.09 + X.08:X.10 + X.09:X.10, family = binomial(link = "logit"),
  data = df1)

Deviance Residuals:
    Min       1Q   Median       3Q      Max
-2.43005  0.00003  0.00312  0.09310  1.89938

Coefficients:
```

	Estimate	Std.Error	z.value	Pr(>|z|)	Code
(Intercept)	3.090371	19.47492	0.159	0.87392	
X.01	-1.78937	0.990729	-1.806	0.0709	.
X.02	2.313062	0.752522	3.074	0.00211	**
X.03	5.590012	3.610579	1.548	0.12157	
X.04	-0.05486	0.017674	-3.104	0.00191	**
X.05	0.11296	0.17171	0.658	0.51063	
X.06	1.483428	0.865458	1.714	0.08652	.
X.07	1.373992	1.56768	0.876	0.38079	
X.08	2.827068	1.428178	1.979	0.04776	*
X.09	31.96744	46.71182	0.684	0.49375	
X.10	-0.2674	1.089954	-0.245	0.8062	
X.01:X.03	-0.04882	0.069773	-0.7	0.48416	
X.01:X.06	-0.0203	0.029078	-0.698	0.48522	
X.01:X.07	-0.10282	0.053465	-1.923	0.05446	.
X.01:X.08	0.017009	0.024288	0.7	0.48374	
X.01:X.09	1.942609	2.126655	0.913	0.361	
X.01:X.10	0.029468	0.038055	0.774	0.43871	
X.03:X.06	0.04179	0.074648	0.56	0.5756	
X.03:X.07	0.142403	0.218422	0.652	0.51443	
X.03:X.08	-0.08221	0.079187	-1.038	0.29917	
X.03:X.09	-13.2495	5.693598	-2.327	0.01996	*
X.03:X.10	0.002034	0.114972	0.018	0.98588	
X.06:X.07	0.094908	0.051953	1.827	0.06773	.
X.06:X.08	0.056842	0.031997	1.777	0.07565	.
X.06:X.09	-1.9362	1.702829	-1.137	0.25552	
X.06:X.10	0.014392	0.042649	0.337	0.73578	
X.07:X.08	-0.05809	0.057591	-1.009	0.31311	
X.07:X.09	2.083368	3.293589	0.633	0.52703	
X.07:X.10	-0.21126	0.084158	-2.51	0.01206	*
X.08:X.09	-8.26828	3.190887	-2.591	0.00956	**
X.08:X.10	-0.11202	0.055535	-2.017	0.04368	*
X.09:X.10	-1.58071	2.275694	-0.695	0.4873	

```
---
Signif. codes: 0 '***' 0.001 '**' 0.01 '*' 0.05 '.' 0.1 ' ' 1
AIC: 113.69
```

Response.B ~ X.01 + X.02 + X.03 + X.04 + X.06 + X.07 + X.08 + X.09 + X.10 + X.01:X.07 + X.03:X.09 + X.06:X.07 + X.06:X.08 + X.06:X.09 + X.07:X.10 + X.08:X.09 + X.08:X.10

Table 15.4 Final terms and parameter estimates—stepwise logistic model

(Code 15.2)

Term	Estimate
(Intercept)	4.369141
X.01	-0.76486
X.02	1.781486
X.03	3.60539
X.04	-0.04623
X.06	0.924242
X.07	1.202791
X.08	2.689647
X.09	10.22816
X.10	-0.11629
X.01:X.07	-0.08569
X.03:X.09	-11.5584
X.06:X.07	0.067759
X.06:X.08	0.046451
X.06:X.09	-1.58571
X.07:X.10	-0.16555
X.08:X.09	-5.9958
X.08:X.10	-0.10876

Thus, the final model had 18 terms (including the intercept), and an AIC of 93.27. Table 15.4 provides the parameter estimates for the final logistic regression model.

Since the observed response is binary, but the model provides predicted probabilities, the predicted values cannot be plotted against the observed. The final stepwise model had 18 parameters. If only significant terms from the full model (including terms that were no necessarily significant by themselves but were in a significant interaction) were included in a reduced model, that reduced model would have 17 terms (including an intercept, which was not significant in the original model). Table 15.5 provides the parameter estimates for the reduced model.

The rule of parsimony would suggest using such a reduced model over the final stepwise model. However, it does not seem as though reducing the number of parameters from 18 to 17 would make much difference. Figure 15.2 shows the probability predictions of the reduced model plotted against the predictions from the final stepwise model.

It is not terribly easy to assess which of these is better. The reduced model had an AIC of 101.34. The final stepwise model had an AIC of 93.27. The reduced model had one fewer parameters than the final stepwise model. The final stepwise model had a lower AIC than the reduced model. Unless there is some reason to suspect that some terms in one model are particularly meaningful or important, there is not much evidence that one of these models is better than the other.

Table 15.5 Terms and parameter estimates—reduced logistic model

(Code 15.2)

Term	Estimate
(Intercept)	-0.993327
X.01	-0.447523
X.02	1.069845
X.04	-0.030299
X.05	0.116097
X.06	0.552258
X.07	0.392858
X.08	1.422849
X.09	0.514702
X.10	-0.114142
X.01:X.07	-0.047901
X.09:X.03	-1.595669
X.06:X.07	0.066434
X.06:X.08	0.010423
X.07:X.10	-0.097588
X.08:X.09	-2.584226
X.08:X.10	-0.057123

(Code 15.2)

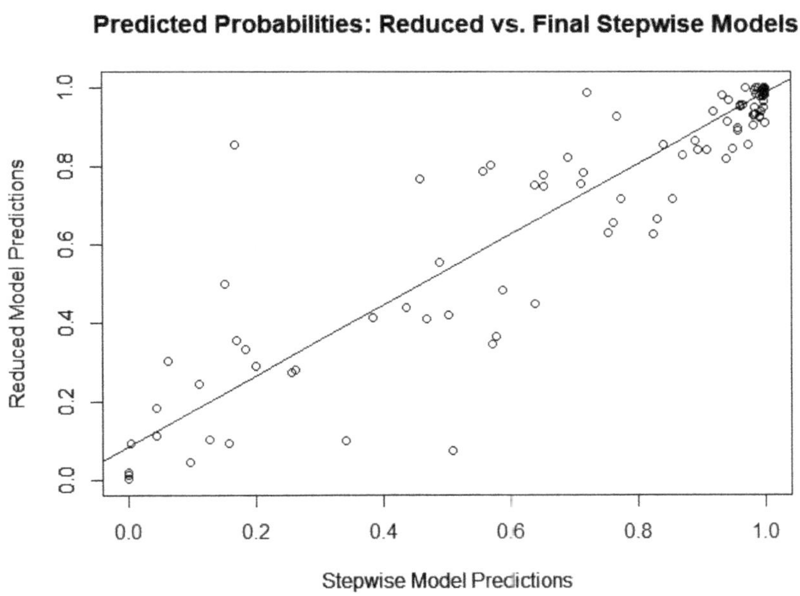

Fig. 15.2 Probability predictions

15.4 Bayesian Model Averaging (BMA) and GLMULTI

Like stepwise regression, Bayesian Model Averaging (BMA) takes an initial "full" model and iteratively finds new models (Hoeting, et al., 1999). BMA uses prior distributions for model parameters to yield a Bayesian posterior model that is an "average" of all the iterations. The final posterior model has all the same terms as the initial model, but the parameter estimates are based on a Bayesian approach rather than ordinary least squares. So the BMA approach ultimately does not eliminate parameters, but provides a different estimate of those parameter values. The posterior values of the parameters is actually an average from the posterior distribution, and depends on the choice of the prior. The R function *bic.glm()* in package BMA implements BMA. The parameter *prior.param* allows the user to specify a prior probability "weight" for each variable in the initial "full" model. The function will then use some internal rules to decide which subset of models (different combinations of the variables in the initial model) should be used for averaging. The choice of models to include depends in part on the *prior.param* input. If that input is not specified, the program will assume a uniform prior probability, which is based on the number of variables in the full model. The values in *prior.param* are not required to sum to 1.

As an example, consider the 10-regressor case with the continuous response variable. Table 15.6 provides all the terms and their associated parameter estimates.

Table 15.6 BMA parameter estimates

(Code 15.2)

Term	Estimate
(Intercept)	6.37E+01
X.01	-2.52E+00
X.02	1.30E-04
X.03	-1.89E+00
X.04	-1.07E-04
X.05	-3.67E-03
X.07	1.43E+00
X.08	-5.00E+00
X.09	-4.04E-01
X.10	-2.09E+00
X.01:X.03	7.22E-03
X.01:X.06	9.30E-02
X.01:X.07	1.16E-03
X.01:X.08	3.15E-05
X.01:X.09	-4.47E-01
X.01:X.10	3.16E-04
X.03:X.06	-5.38E-02
X.03:X.07	-6.33E-02
X.03:X.08	-2.55E-03
X.03:X.09	-4.46E-01
X.03:X.10	-1.55E-03
X.06:X.07	-2.62E-01
X.06:X.08	1.18E-01
X.06:X.10	1.54E-01
X.07:X.08	-3.39E-04
X.07:X.09	-1.58E-02
X.07:X.10	3.63E-02
X.08:X.09	3.53E-02
X.08:X.10	5.66E-03
X.09:X.10	-1.33E-02

(Code 15.2)

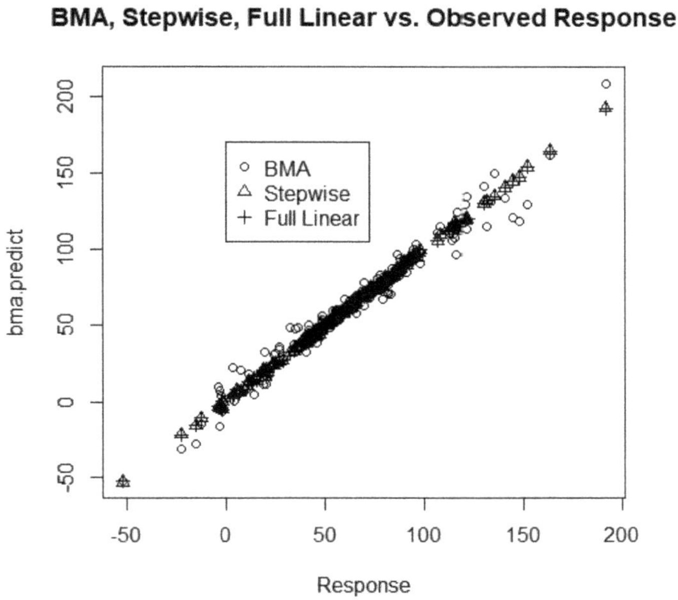

Fig. 15.3 Predicted vs. observed response: BMA, stepwise, and full models

Figure 15.3 shows a plot of predicted versus observed response values for BMA, the final stepwise model, and the full (unreduced) linear model.

15.5 GLMULTI

Model averaging can be done in many ways (Buckland, et al., 1997). There is a function in R package **glmulti** called *glmulti*() (Calcagno, 2010). This function is similar to stepwise regression, but uses a different process to select models. Furthermore, the "full" model need not be specified. Rather, a formula with all the main effects to be included is specified, and then an input parameter tells *glmulti*() whether to consider two-way interactions or only main effects in selecting a model. The terms and associated estimates of the "best" model (lowest AIC) for the 10-regressor example are given in Table 15.7 The final output will have average predictions over a set of "best" models.

Figure 15.4 is like Fig. 15.3, but with the GLMULTI "average" predictions also plotted.

Table 15.7 GLMULTI parameter estimates—"Best" model

(Code 15.2)

Term	Estimate
(Intercept)	-9.11E+00
X.01	-1.07E+00
X.03	-2.74E+00
X.05	-2.24E-01
X.06	4.91E+00
X.07	-2.21E+00
X.08	-3.22E+00
X.10	9.65E-02
X.01:X.02	9.33E-03
X.05:X.06	7.16E-03
X.03:X.07	-5.47E-02
X.01:X.08	-7.97E-03
X.04:X.08	-6.94E-04
X.05:X.08	8.32E-03
X.01:X.09	-3.21E-01
X.04:X.09	1.61E-02
X.05:X.09	1.89E-01
X.07:X.10	1.25E-02

(Code 15.2)

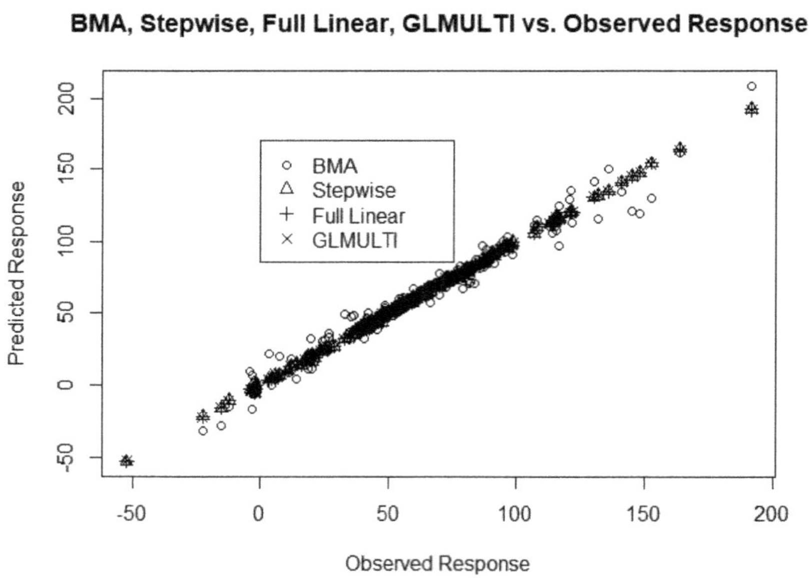

Fig. 15.4 Predicted vs. observed response: BMA, stepwise, full, and GLMULTI models

15.6 Summary

Selecting a model, even if only linear models are being considered, is not trivial, especially when many potentially meaningful regressors are available. Possibly the oldest methodology is referred to as stepwise regression. The stepwise approach selects terms to include or exclude based on some numerical criterion, such as AIC. A sort of "manual" approach might be to first include all possible terms in a model, and then reduce the model to only those terms with significant parameters (i.e., parameters that are significantly different from 0). There are other methods, including Bayesian model Averaging and GLMULTI. These methods use an "average" model to yield a final prediction. The modeler would then need to determine whether the terms included in the model seem to explain the response predictions in a manner that either fits with other scientific knowledge or provide some reasonable hypotheses about the relationship between the terms and the response. There is great temptation to simply rely on function-fitting methods to provide predictions, without providing any insight as to the relationships between regressors and responses.

15.7 Points to Ponder

1. What advantage does AIC have over R^2 as a measure of model adequacy?
2. Is there any benefit of the stepwise approach to model reduction by eliminating non-significant regressors?
3. Can you envision conditions under which Bayesian Model Averaging might be very advantageous?

References

Buckland, S. T., Burnham, K. P., & Augustin, N. H. (1997). Model selection: An integral part of inference. *Biometrics, 53*(2), 603–618.

Calcagno, V. (2010). 127.0.0.1:26044/library/glmulti/html/glmulti.html . R: Automated model selection and multimodel inference with...

Draper, N. R., & Smith, H. (1981). *Applied regression analysis* (2nd ed.). Wiley.

Hoeting, J., Madigan, D., Raftery, A. E., & Volinsky, C. T. (1999). Bayesian model averaging: A tutorial. *Statistical Science, 14*(4), 382–417.

Chapter 16
Multivariate Analysis

In multiple linear regression, models have multiple regressors. However, there was only one response variable. Multivariate analysis involves multiple response variables simultaneously. In general, the responses are correlated with each other. Commonly, the objective is to be able to categorize individuals based on the values of the responses. In many cases, there are several groups or classes of individuals, and the objective is to use the response variables that would first define some boundaries of those variables so that individuals can be discriminated into their groups.

We will concentrate on the objective classification. The classification will further fall into two types, namely when in the data set it is known a priori into which category each individual actually belongs, and it is desired to build a decision function that will allow for classifying an "unknown" individual simply by having all the variables measured or observed. The second type is where there is no a priori set of classes or categories, and the researcher desires to discover whether or not the data themselves suggest any partitioning of individuals into some heretofore unknown set of categories.

An important notion for multidimensional data is that of eigenvectors (also called characteristic vectors). These represent a new set of dimensions, so that each unit can be mapped into a "space" where the axes are orthogonal (at right angles) to each other. In the case of data analyses, the new dimensions are based on the degree to which variation in the whole data set occurs. The new dimensions are the eigen vectors of the correlation matrix for the multiple variables and are called principal components. The number of principal components is usually the number of variables, but sometimes the data are such that one component appears more than once. The covariance matrix could also be used to compute principal components, but the correlation matrix is scaled to that the entries in the matrix are unitless.

Supplementary Information The online version contains supplementary material available at https://doi.org/10.1007/978-3-031-84479-9_16. The videos can be accessed individually by clicking the DOI link in the accompanying figure caption or by scanning this link with the SN More Media App.

S. A. Pardo, *Beyond the T-Test*, https://doi.org/10.1007/978-3-031-84479-9_16

16.1 Prior Category Assignment

Suppose in a set of n experimental units, each unit belongs to exactly one of G groups, so that the groups have a null intersection. That is, membership in one group is mutually exclusive of all the other groups. Furthermore, for each unit, several random variables, $X_1, X_2, \ldots, X_m$, are observed or measured. The question is whether knowing the values of an individual's variables, it is possible to identify the group to which the individual belongs. Given that the variables that characterize an individual have a random component, at best individuals could be assigned a probability of belonging to any particular group or category. At the least, group membership might be determined on some kind of average condition.

An older method for generating a function that assigns individuals to groups is called linear discriminant analysis (Anderson, 1958). For each individual, the set of values for the characterizing variables form an m-tuple vector:

$$\left[x_1, x_2, \ldots, x_m\right]$$

A sample covariance matrix gives estimates of the covariance between every pair of the variables. That is, the sample covariance between X_i and X_j is:

$$s_{ij} = \frac{1}{n}\sum_{k=1}^{n}\left(x_{ki} - \overline{x}_i\right)\left(x_{kj} - \overline{x}_j\right)$$

So x_{ki} is the observation/measurement of variable X_i on experimental unit k, and $\overline{x}_i$ is the average value of X_i, averaged over all the experimental units. So the matrix:

$$\boldsymbol{S} = \begin{bmatrix} s_{11} & \cdots & s_{1m} \\ \vdots & \ddots & \vdots \\ s_{m1} & \cdots & s_{mm} \end{bmatrix}$$

is called the sample covariance matrix. Its dimensions are $m \times m$, and it is symmetric, i.e.,

$$s_{ij} = s_{ji}.$$

Individual k can be characterized by a vector of m values:

$$\boldsymbol{x}_k = \begin{bmatrix} x_{k1} \\ \vdots \\ x_{km} \end{bmatrix}$$

Recall that the individuals are known to belong to exactly one of G groups. A vector of averages can be constructed for each group:

$$\overline{\boldsymbol{x}}_g = \begin{bmatrix} \overline{x}_{g1} \\ \vdots \\ \overline{x}_{gm} \end{bmatrix}$$

That is, the components of this vector, $\overline{\boldsymbol{x}}_{gi}$, are the average values of variable i in group g. Then, for each group $g = 1, G$, form a function of an individual's data vector:

$$d_g(\boldsymbol{x}_k) = (\boldsymbol{x}_k - \overline{\boldsymbol{x}}_g)^T \boldsymbol{S}^{-1} (\boldsymbol{x}_k - \overline{x}_g), g = 1, G$$

The notation: v^T is the transpose of vector v. Vectors in this work are assumed to be columns unless otherwise stated. So if the vector v is an $m \times 1$ column vector, v^T would be a row vector of dimensions $1 \times m$.

The form of this function is called Mahalanobis' distance (Anderson, ref. cit.). The idea of linear discriminant analysis is to compute all the d_g for each individual, and the individual's "predicted" group would be the one in which this distance is minimized. Of course, an individual whose data were not used to compute the mean vectors and the covariance matrix, and for whom group membership is unknown, can be classified by computing the d_g using the new individual's data. Thus, the set of individuals whose data were used to compute the discriminant functions, d_g, are referred to as the "training set." Another set of individuals, for whom group membership is known, but whose data were not used, can form a "test" set. That is, the discriminant functions can be used to predict group membership for each individual in the test set, and the prediction can be compared to their "known" group membership. The percentage of test set individuals for whom the predicted group matches the actual group is a measure of goodness for the discriminant functions.

Mahalanobis' distance is a fairly simple way to form discriminant functions. Sir Ronald Fisher (Fisher, 1936) had a slightly more sophisticated approach. He decomposed the covariance matrix into two terms, between-groups and within-groups. The within-group term first creates a set of $m \times m$ matrices for the form:

$$\boldsymbol{W}_{kg} = (\boldsymbol{x}_{kg} - \overline{\boldsymbol{x}}_g)(\boldsymbol{x}_{kg} - \overline{\boldsymbol{x}}_g)^T$$

The $\boldsymbol{x}_{kg}$ is the vector of observations for the kth individual in group g. Its dimensions are $m \times 1$. The matrix $\boldsymbol{W}_{kg}$ has dimensions $m \times m$. Each individual has its own $\boldsymbol{W}_{kg}$ matrix. They are summed to produce the "total" within-group part of the covariance:

$$\boldsymbol{W} = \sum_k \sum_g \boldsymbol{W}_{kg}$$

Each group has its own matrix that expresses how the group differences from the sort of centroid of all the groups:

$$\boldsymbol{B}_g = (\overline{\boldsymbol{x}}_g - \overline{\boldsymbol{x}})(\overline{\boldsymbol{x}}_g - \overline{\boldsymbol{x}})^T$$

The $\bar{\boldsymbol{x}}$ is the average vector, averaged over all individuals regardless of their group. Each matrix $\boldsymbol{B}g$ also has dimensions $m \times m$. The overall "between group" part of the covariance is:

$$\boldsymbol{B} = \sum_{g} \boldsymbol{B}_g$$

Both $\boldsymbol{W}$ and $\boldsymbol{B}$ are square $m \times m$ matrices, so that they and their inverses (provided they exist) are conformable (i.e., they can be multiplied together). Fisher's discriminant functions are the inner products of an individual's data vector, $\boldsymbol{x}_k$, and each eigenvector of the matrix:

$$W^{-1}B$$

Often the eigenvectors are "normalized" by multiplying each of them by the reciprocal of their respective magnitudes. So the gth discriminant function can be expressed as:

$$d_g\left(\boldsymbol{x}_k\right) = \boldsymbol{x}_k \cdot \in_{\boldsymbol{g}}$$

where $\boldsymbol{\epsilon}_{\boldsymbol{g}}$ is the gth eigenvector.

The number of discriminant functions is based on the eigenvectors of $\boldsymbol{W}^{-1}\boldsymbol{B}$ corresponding to the G largest eigenvalues of that matrix. Note that the number of groups, G, must be less than or equal to the number of variables, m.

As in the case of discriminant functions based on Mahalanobis' distance, an individual's predicted group is based on the discriminant function with the smallest value.

Once the discriminant functions have been formed, they can be used to classify a new individual into one of the known groups or classes.

As an example, suppose $n = 100$ individual units were assigned to one of four groups. Linear discriminant analysis (LDA) can be used to find a classification for new individuals to assign them to one of the four groups.

Table 16.1 provides the discriminant functions for these data. There are three functions, one less than the number of groups.

Table 16.1 Linear discriminant functions (Coefficients of Eigenvectors)

(Code 16.1)

Coefficients of linear discriminants:

Variable	LD1	LD2	LD3
X0	3.40E-01	-1.53E-01	2.36E-02
X1	-2.09E-01	2.04E-01	5.95E-02
X2	-5.41E-02	2.77E-01	-5.08E-02
X3	3.80E-01	-1.00E+00	1.26E+00
X4	5.57E-02	-5.65E-02	2.36E-02
X5	-1.92E-06	-1.16E-06	1.81E-06
X6	1.36E-02	-2.14E-02	1.72E-02
X7	4.29E+00	1.02E-01	-5.10E+00
X8	6.23E-03	-7.18E-02	4.63E-02
X9	1.03E-01	1.36E-02	7.63E-02

(Code 16.1)

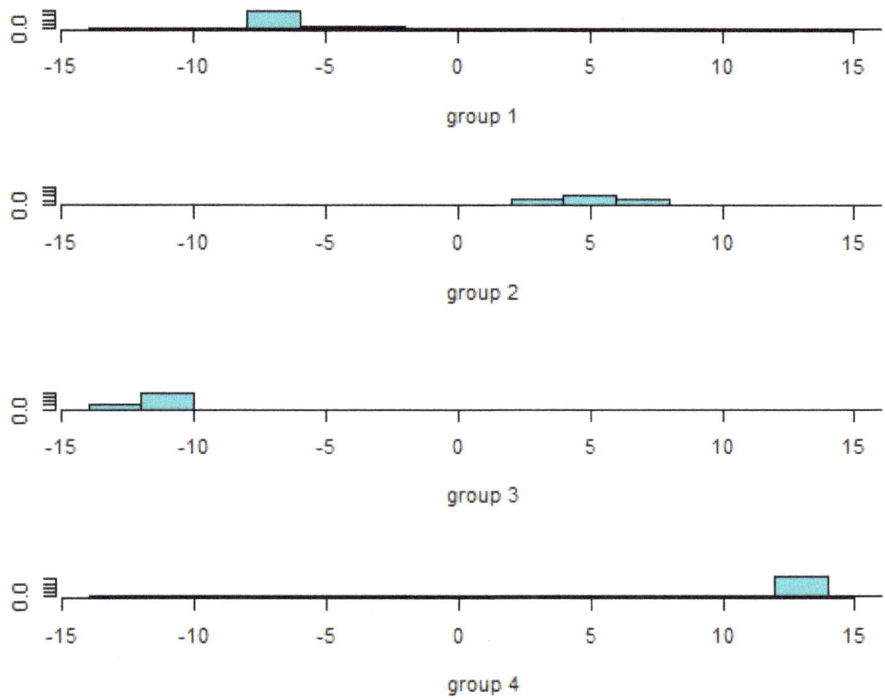

Fig. 16.1 Histograms of LD1 values by group

Figure 16.1 shows the histograms of the value of LD1 for the data in the four a priori groupings.

16.2 No Prior Category Assignment

How to find groups of individual units, where no groupings existed before? It is a little like deep space exploration; you have a multidimensional space, and you find lots of individuals (maybe planets, maybe stars, maybe alien life forms). You want to know if they can be categorized. You make many measurements of various types on each individual you meet. Now what? One set of methods for finding groupings in a multidimensional set of data is called cluster analysis (Everitt, et al., 2011). The basic idea is to first have some measure, or metric, for distance between two individuals, in terms of the multidimensional space described by all the variables you have measured or observed. Then there are many different methods or algorithms that use this distance to decide which individuals are "close" to each other and which are far apart. In the language of machine learning, these methods are all called "unsupervised" learning, since there is no mapping between inputs and known outputs (categories). Rather, the algorithm attempts to define the categories. While cluster analysis is a broad topic, we will focus on one subset of methods referred to as "*k* means." That is, the analyst decides that there must be *k* groups, of unknown definition, and asks the algorithm to determine which units belong to each group. The groups are mutually exclusive; an individual can belong to exactly one group. Once the algorithm has assigned each individual to a group, it is up to the researcher to decide what makes these groups interesting.

The primary problem with *k* means algorithms is that the analyst must guess at how many groups there might be. One tool for helping to solve this problem is to try several possible numbers of groups, and then find the number that beyond which there does not seem to be any improvement is some distance measure. So if $\boldsymbol{x}_{ij}$ represents a vector of measurements for the *i*th experimental unit in the *j*th group then the idea is to find the number of groups, *k*, such that:

$$d = \sum_{j=1}^{k} \sum_{i=1}^{n_j} \left(\boldsymbol{x}_{ij} - \bar{\boldsymbol{x}}_j \right)^2$$

is not only a minimum, but does not appear to get smaller by using a larger number of groups. The distance, *d*, is referred to as the within-cluster sums-of-squares.

(Code 16.1)

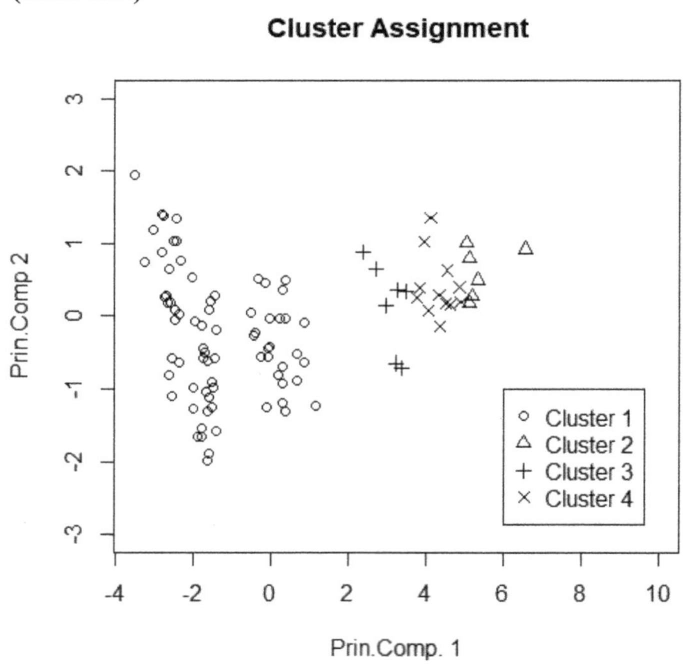

Fig. 16.2 Cluster assignment

Table 16.2 Correspondence between group and cluster

(Code 16.1)

	cluster.id			
Group	**1**	**2**	**3**	**4**
1	25	0	0	0
2	25	0	0	0
3	25	0	0	0
4	0	6	7	12

The clusters are illustrated in Fig. 16.2 by plotting the values of the first and second principal components of the correlation matrix and symbolizing each unit by its corresponding cluster. In this plot, the k means algorithm was told to try to find four clusters.

The units were initially assigned to one of four groups. Table 16.2 provides the correspondence between group and cluster.

So all the individuals in Groups 1, 2, and 3 fell into cluster 1. The units in Group 4 were split between clusters 2, 3, and 4. It is possible there are only two groups. It is equally likely that most individuals fall into one group, but there are those who fall into one of three other groups. Only deeper understanding of the physics or biology or sociology or psychology of the variables can help determine which of those possibilities makes the most sense.

16.3 Summary

Unlike multiple regression, multivariate analyses involve more than one response, where those response variables are correlated. The primary goal of multivariate analysis as discussed here is to categorize individuals into one of several groups. In the case of linear discriminant analysis, those groups are assigned to individuals prior to any data being gathered. In the case of cluster analysis, the data are used to define groups.

16.4 Points to Ponder

1. Why is multiple linear regression not considered multivariate analysis?
2. What is the primary goal of linear discriminant analysis? How is it similar or different to cluster analysis?

References

Anderson, T. W. (1958). *An introduction to multivariate statistical analysis*. Wiley.

Everitt, B. S., Landau, S., Leese, M., & Stahl, D. (2011). *Cluster analysis* (5th ed.). Wiley.

Fisher, R. A., (1936). *The Use of Multiple Measurements in Taxonomic Problems, Annals of Eugenics*, 7, 179–188.

Chapter 17
Bayesian Methods—Markov Chain Montel Carlo Approach

17.1 The Basic Idea of Bayesian Inference

The primary notion of Bayesian analyses is that the analyst has some information about the parameters of interest prior to gathering data. The nature of this information is probabilistic. That is, the analyst has some idea of the range of possible values for the (unknown) parameters, and the information is contained in a probability distribution. This is not to necessarily say that parameters do not have "true" values; rather, we are quantifying our knowledge of the parameter values in terms of uncertainty. Thus the analyst gathers data, then using the likelihood function (that is admittedly a model) the analyst will "update" the information contained in the prior distribution using Bayes' Theorem, to create the "posterior" distribution, as described in Chap. 1. The formula for computing the posterior density function, given the prior density and the likelihood function, is:

$$f_{\text{posterior}}\left(\tilde{\mu}|x_1,x_2,\ldots,x_n\right)=\frac{f_{\text{prior}}\left(\tilde{\mu}\right)L\left[x_1,x_2,\ldots,x_n|\tilde{\mu}\right]}{\int_{-\infty}^{+\infty}f_{\text{prior}}\left(\xi\right)L\left[x_1,x_2,\ldots,x_n|\xi\right]d\xi}$$

The function $L\left[x_1,x_2,\ldots,x_n|\tilde{\mu}\right]$ is the likelihood of obtaining the data $x_1, x_2, \ldots, x_n$ given that the parameter (or vector of parameters) is $\tilde{\mu}$. The likelihood function for a random sample $x_1, x_2, \ldots, x_n$, is:

$$L\left[x_1,x_2,\ldots,x_n|\tilde{\mu}\right]=\prod_{i=1}^{n}f_x\left(x\,|\tilde{\mu}\right)$$

That is, $f_x\left(x_i|\tilde{\mu}\right)$ is a density function conditioned on $\tilde{\mu}$ and is the same function for all the x_i.

S. A. Pardo, *Beyond the T-Test*, https://doi.org/10.1007/978-3-031-84479-9_17

So Bayesian analyses, i.e., updating the prior, depends on the ability to compute the integral:

$$\int_{-\infty}^{+\infty} f_{\text{prior}}(\xi) L[x_1, x_2, \ldots, x_n | \xi] d\xi$$

There are certain combinations of prior densities and likelihood functions that all that integral to be evaluated in a closed form, and in fact result in a posterior density of the same family as the prior. Such combinations are referred to as "conjugate pairs" (Gelman, et al., 1997). Such pairs were at one time the only practical models for which Bayes' Theorem could be applied.

17.2 The Markov Chain Monte Carlo (MCMC) Method

While various numerical methods might be applied to the evaluation of the integral for non-conjugate pairs, such methods are not necessarily practical, due to limitations in computing. In more recent times, a methodology referred to as the Markov Chain Monte Carlo (MCMC) approach, simplified the computations. An algorithm called the Metropolis-Hastings algorithm, made implementing MCMC fairly straightforward. This is how it works:

1. Initially and arbitrarily choose a value for $\tilde{\mu}$, call it $\tilde{\mu}_0$.
2. Randomly choose another value for $\tilde{\mu}$, call it $\tilde{\mu}_1$. You can use any method you like for randomly generating a value (or a vector) $\tilde{\mu}_1$. One way would be to randomly generate a value (or a vector, if there are multiple parameters) from the prior distribution $g(\tilde{\mu})$
3. Compute the ratio:

$$R = \frac{g(\tilde{\mu}_1) f_X(x|\tilde{\mu}_1)}{g(\tilde{\mu}_0) f_X(x|\tilde{\mu}_0)}$$

   $f_X(x|\tilde{\mu})$ is the likelihood function.
4. Randomly generate a number between 0 and 1, and call it u. If $R \geq u$, then the new value for $\tilde{\mu}$ is $\tilde{\mu}_1$. Store it in the $\tilde{\mu}$ list. Otherwise, store $\tilde{\mu}_0$.
5. Assign to $\tilde{\mu}_0$ the newly stored value.
6. Repeat the process n times from step #2.

The idea is that the ratio, R, is actually the ratio of two posterior density valuations. Since both the numerator and denominator are of the form:

$$\frac{g(\tilde{\mu}) f[x|\tilde{\mu}]}{\int_{-\infty}^{+\infty} g(\xi) f[f|\xi] d\xi}$$

The integral cancels, so that evaluating it is unnecessary. Thus, as long at the product:

$$g(\gamma)f(x|\gamma)$$

is easily computed, the forms of $g(.)$ and $f(.)$ are far less restricted, i.e., they do not have to be a conjugate pair.

Once the posterior distribution is found, a lower and upper percentile of the posterior can be used as a sort of interval estimate, called a credible interval, for the parameter of interest. Typically, the 2.5th and 97.5th percentiles are used to for a 95% credible interval. The interpretation of such intervals is that we believe parameter value falls in the interval with probability determined by the tail probabilities of the interval limits (e.g., if the lower limit is the 2.5th percentile and the upper limit is the 97.5th percentile, then the probability content is 97.5%–2.5% = 95%). In the following two examples, results from an MCMC algorithm are compared to those obtained using the conjugacy of the chosen prior and likelihood functions.

17.3 Beta-Binomial

As an example consider a binomial experiment, where x "successes" are observed out of n independent and identically distributed Bernoulli trials (a Bernoulli trial is an experiment where only 1 of 2 possible outcomes can occur, where one type of outcome is considered a "success"). The probability of a success on any given trial, p, is the unknown parameter. Suppose that we postulate a prior for p, and that it is of the Beta distribution family. The Beta and Binomial distributions form a conjugate pair. We will compare results from the calculation of posterior meta-parameters using the convenience of conjugacy, to those derived from MCMC.

The prior density can be described as:

$$g(p|\alpha,\beta)=\frac{\Gamma(\alpha+\beta)}{\Gamma(\alpha)\Gamma(\beta)}p^{\alpha-1}(1-p)^{\beta-1}$$

The "meta-parameters," α and β, are values that define the exact shape of the prior density for p.

The expected value of the Beta is:

$$E(p)=\frac{\alpha}{\alpha+\beta}$$

The likelihood function is the Binomial mass function:

$$f(y|p)=\binom{n}{y}p^{y}(1-p)^{n-y}$$

So, applying Bayes' Theorem would yield a posterior density, call it *g'*, that would have the form:

$$g'(p|\alpha',\beta') = \frac{\Gamma(\alpha'+\beta')}{\Gamma(\alpha')\Gamma(\beta')} p^{\alpha'-1}(1-p)^{\beta'-1}$$

Suppose that an experiment was performed in which $n = 50$ individuals were determined to be either "good" or "not good." Furthermore, suppose that of those 50, $y = 40$ were "good," so that.

A point estimate of the binomial parameter, *p*, is:

$$\widehat{p} = \frac{y}{n} = \frac{40}{50} = 0.80$$

The analysts chooses to use a Bayesian analysis, with a Beta prior having "meta" parameters:

$$\alpha = 1.0(\text{shape})$$

$$\beta = 0.25(\text{scale})$$

As to why those prior "meta" parameters, in theory they provide a prior distribution that captures the analyst's guess about the range of possible values for *p*. In this case, the prior expected value is:

$$E(p) = \frac{\alpha}{\alpha+\beta} = \frac{1}{1+0.25} = 0.80$$

This is the value of the point estimate from the observed data.

Figure 17.1 shows the prior distribution of the parameter *p*.

After 2000 iterations of the MCMC algorithm, the posterior distribution (histogram) of *p* is computed. It is shown in Fig. 17.2.

The MCMC posterior 95% credible interval is (0.6970, 0.8932). The median of the MCMC posterior data is 0.8065. Since the Beta and Binomial distributions are a conjugate pair, the posterior meta-parameters can be computed:

$$\alpha' = \alpha + y = 1 + 40$$

$$\beta' = \beta + n - y = 0.25 + 50 - 40$$

(Code 17.1)

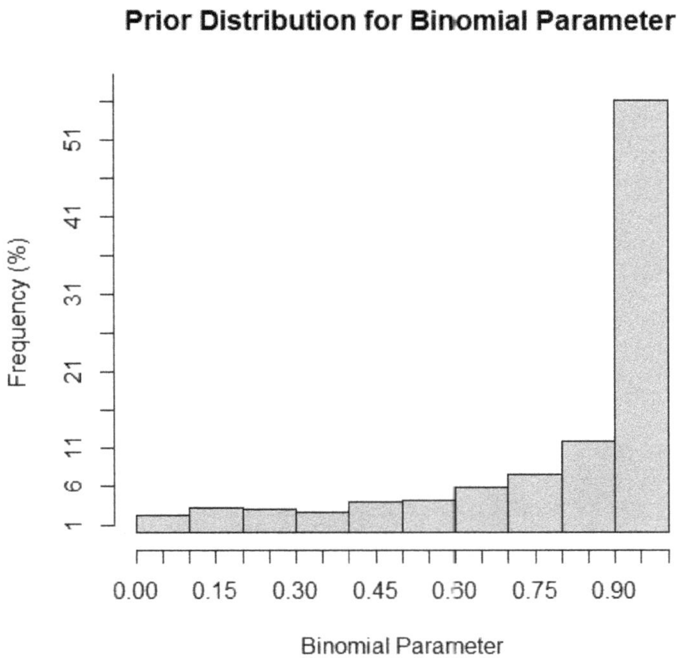

Fig. 17.1 Prior distribution of binomial parameter p

The conjugate posterior is also a Beta distribution. The conjugate posterior 95% credible interval is (0.6813, 0.8965). The conjugate posterior median is 0.8039.

A frequentist 95% confidence interval for the probability of "success" is (0.6628, 0.8997).

Table 17.1 shows these intervals and measures of central tendency.

(Code 17.1)

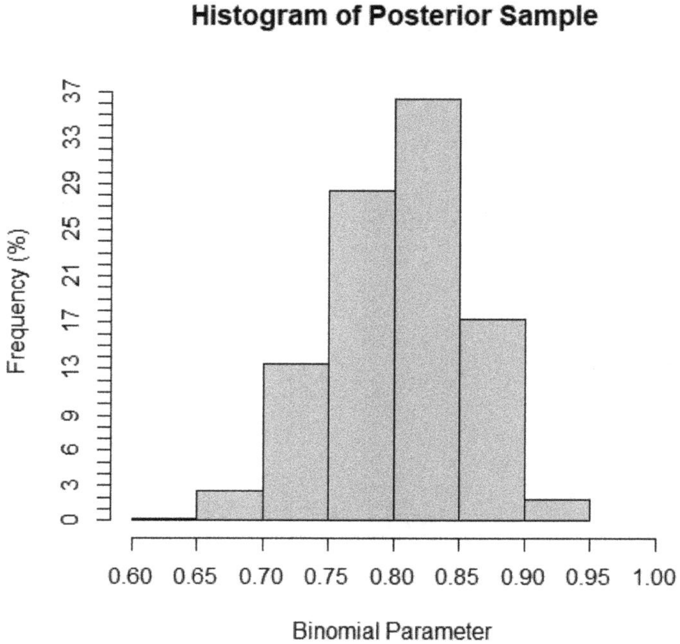

Fig. 17.2 MCMC posterior distribution of $\boldsymbol{p}$

Table 17.1 Credible and confidence intervals—beta-binomial

(Code 17.1)

	MCMC	Conjugate Pair	Frequentist
LL*	0.6970	0.6813	0.6628
UL*	0.8932	0.8965	0.8997
Centraility†	0.8065	0.8039	0.8000

*(LL,UL) = 95% interval (credible or confidence)
†median or mean

17.4 Normal-Normal

Another conjugate pair is a normal prior for μ and a normal likelihood function for the sample statistic $\overline{x}$. Assume that the standard deviation, σ, is actually known. The meta-parameters for μ are γ and τ, so the prior density is:

$$g(\mu|\gamma,\tau)=\frac{1}{\sqrt{2\pi\tau}}e^{-\frac{1}{2}\left(\frac{\mu-\gamma}{\tau}\right)^2}$$

(Code 17.2)

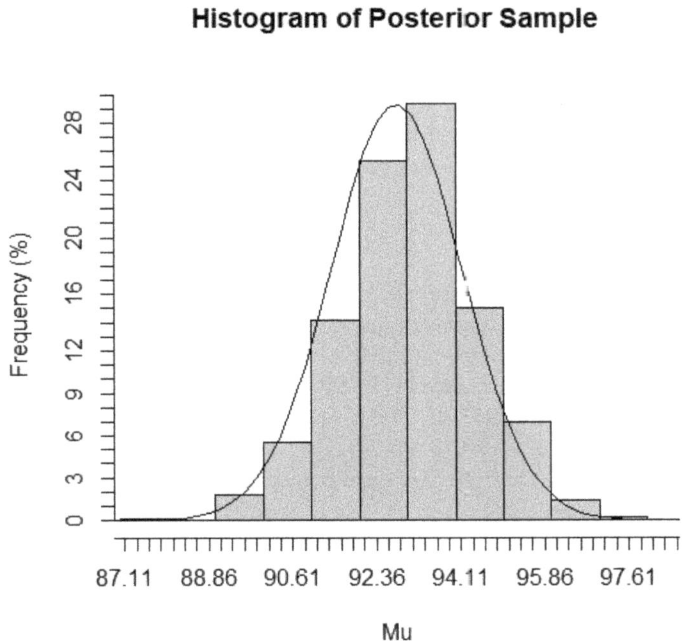

Fig. 17.3 MCMC posterior distribution of μ

The likelihood is in terms of $\bar{x}$:

$$f(\bar{x}|\mu,\sigma,n) = \frac{\sqrt{n}}{\sqrt{2\pi}\sigma} e^{-\frac{1}{2}\left(\frac{\bar{x}-\mu}{\sigma/\sqrt{n}}\right)^2}$$

The posterior meta-parameters are:

$$\gamma' = \frac{n\tau^2\bar{x} + \sigma^2\gamma}{n\tau^2 + \sigma^2}$$

and

$$\tau' = \sqrt{\frac{\sigma^2\tau^2}{n\tau^2 + \sigma^2}}$$

Suppose that in a sample of $n = 50$, $\bar{x} = 93$. The population standard deviation, σ, is known to be 10. The prior parameters were:

$$\gamma = 90$$

Table 17.2 Credible and confidence intervals—normal-normal

(Code 17.2)

	MCMC	Conjugate Pair	Frequentist
LL*	90.22	90.11	90.23
UL*	95.68	95.44	95.77
Centrailityt	93.97	92.78	93.00

*(LL,UL) = 95% interval (credible or confidence)

†median or mean

$$\tau = 5$$

Figure 17.3 shows the histogram of the posterior MCMC samples, together with the posterior density based on the conjugate pair, Normal-Normal.

The 95% credible interval from the MCMC process is (90.22, 95.68). The median of the MCMC posterior distribution is 93.97. Based on conjugacy, the 95% credible interval for μ was (90.11, 95.44). The posterior median was 92.78. A frequentist 95% confidence interval is (90.23, 95.77).

Table 17.2 provides these intervals and measures of central tendency.

17.5 Summary

Bayesian methods allow the experimenter to incorporate into analyses some information about parameters of interest before any experimentation is performed. This prior information might be derived from previous experimentation, from some knowledge of the data-generating process, or from hypotheses. The notion is that the parameter of interest is unknown, but there is some idea about the range of values in which the truth is expected to be. Once data are gathered, they are then used together with the prior knowledge to "update" that knowledge. The prior knowledge comes in the form of a density function for the parameter, and the data are incorporated using a likelihood function. Bayesian analyses requires integration that could be mathematically intractable. For a long time, only particular combinations of prior densities and likelihood functions could be used practically. Such combinations were referred to as conjugate pairs. A simulation-based approach, Markov Chain Monte Carlo (MCMC) methods, has made Bayesian methodology usable in a wider array of situations.

17.6 Points to Ponder

1. How to conjugate pair models simplify Bayesian analyses?
2. How could you use a Bayesian approach to analyze data that are obtained sequentially, perhaps over time?

Reference

Gelman, A., Carlin, J. B., Stern, H. S., & Rubin, D. B. (1997). *Bayesian data analysis*. Chapman & Hall/CRC.

Chapter 18
Machine Learning and Data-Intensive Methods

18.1 Do Machines Learn?

Machine learning methods are, loosely speaking, a collection of model-building methods where an algorithm determines the model used for prediction. An algorithm is a procedure or process, usually iterative in nature, that has some a priori rule indicating when the iteration is complete. The iteration must occur in such a way as to update the inputs, which in turn changes the value of some measure of desirability, called the objective function. Usually an algorithm requires some inputs, such as starting conditions. Furthermore, the rule is specified in such a way as to guarantee the process will complete. Sometimes the completion of the algorithm is called "convergence". So, for example, consider the process of making toast. The bread is sliced and then placed into the toaster. The heating element is then activated, and will remain activated until the preset time has expired or there is manual intervention, whichever comes first. One might refer to the ending of the process when a piece of bread is successfully toasted as "convergence." The toasting process is not exactly iterative, inasmuch as time is a continuous variable. Nevertheless, it is an algorithm. If there were no timer or intervention, then the toaster would continue to cook the toast, ostensibly forever, although most likely the toaster would break down or burst into flames, thus stopping the process. If one checked at regular time intervals to determine if the bread was sufficiently toasted, the process would be iterative. To be an algorithm, there must be absolute certainty that the process will end.

On occasion, software that implements an algorithm doesn't work right, often giving the user a message such as "algorithm failed to converge." The programmers

---

Supplementary Information The online version contains supplementary material available at https://doi.org/10.1007/978-3-031-84479-9_18. The videos can be accessed individually by clicking the DOI link in the accompanying figure caption or by scanning this link with the SN More Media App.

S. A. Pardo, *Beyond the T-Test*, https://doi.org/10.1007/978-3-031-84479-9_18

have put into their code a failsafe mechanism that would not allow any process to continue infinitely. So, on occasion, the numerical precision of the computer is such that it cannot proceed to get closer to achieving the stopping rule. Thus, the program will artificially end the process, and tell the user that something did not go according to plan. Since all digital computers represent continuous, "floating point" numbers by a finite number of bits (a bit is a numeric value that can only be translated as either 0 or 1), there is actually a smallest number that the computer can represent, and a largest number. When programs give the user a message such as "algorithm failed to converge," it is often due to some computation that comes very close to the largest number the computer can represent or due to division by a number smaller than the smallest number the computer can represent (effectively, division by zero). Sometimes the failure to converge is due to the starting or initial conditions, or some other inputs that are outside the range of inputs anticipated by the programmer.

Algorithms are based on something called objective functions. This is some quantitative formula that is dependent on inputs, and whose value gets altered through the course of the algorithmic process. The stopping rule usually involves checking the objective function's value periodically, and stopping when the objective has not changed "substantially" since the last time it was checked. The inputs to the objective function are altered at each stage of the process, hopefully making the objective function's value change to some "optimal" point.

Technically speaking, least squares multiple regression is an algorithm, but the model is chosen by the analyst, and not by a computer program. Machine learning algorithms are sometimes referred to as unsupervised learning, in that the decision about model details is determined by the program, and not by the human analyst. Even Bayesian sequential analysis, while employing an algorithm, still requires human specification of prior distributions and likelihood functions. We will explore three types of machine learning algorithms. Even these are not entirely free of human intervention, but they do rely more heavily than some older methods on algorithmic determination of predictive models.

18.2 Artificial Neural Networks

The original idea behind the artificial neural network (Kuhn & Johnson, 2013) was that it should mimic the way in which biological neurons transmit data so that the organism "learns." As it turns out, artificial neural networks do not actually mimic biological neurons in any real way. However, they are a useful tool for building a predictive model when the relationships between response and predictors are not understood. The way they work is by making guesses at how to relate a particular set of predictor or regressor variables with responses using a "training set" of data. Then, a test set can be used to see just how good the guess was. The tool is called a neural network, because there are layers through which the information moves. The outmost layers are the inputs and the responses. Between those layers are the nerve cells. Each cell will get "stimulated" by each input, and then produce some sort of

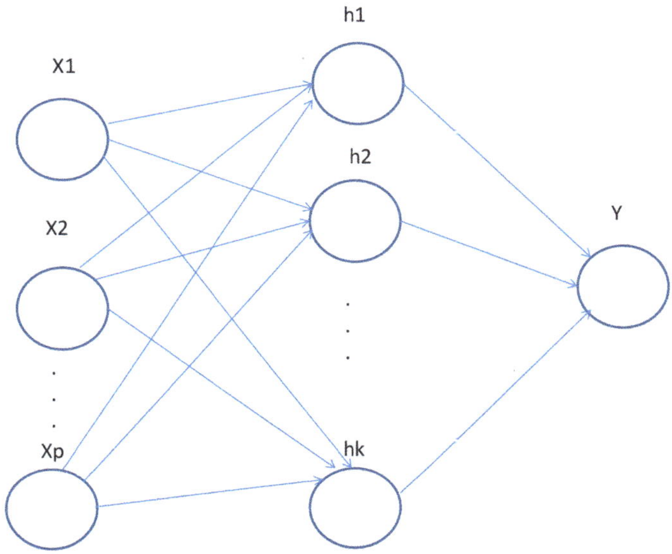

Fig. 18.1 A conceptual neural network

"output," that can then feed into the next layer, etc., until the final neural layer produces a "response." The internal layers of nerve cells are called "hidden" layers. Most applications seem to do best with a single hidden layer. Each nerve cell (neuron) consists of a little function with some parameter that is estimated using the training data. In other words, the algorithm finds the "best" values of the parameters so that the predicted responses most closely match the observed responses. Fig. 18.1 illustrates the concept of the artificial neural network (referred to from here on as neural networks).

Neural networks can be used in two ways. First, they can be used to classify unknown individuals into one of several a priori defined classes, as in the case of linear discriminant analyses. They can also be used as multiple regression models, to compute a predictive equation for continuously values responses.

The little functions within each neuron are often the logistic "S" curve function (usually most appropriate when the neural net is used for classification):

$$h(x) = \frac{1}{1 + e^{-\beta^T x}}$$

However, the functions in the hidden layer(s) can be simply linear:

$$h(\boldsymbol{x}) = \boldsymbol{\beta}^T \boldsymbol{x}$$

There is no actual restriction on the forms of the $h(\boldsymbol{x})$ functions. However, the S-shaped curves and linear curves seem to be the most common choices. The variable $\boldsymbol{x}$ is a vector of inputs from the previous layer, and $\boldsymbol{\beta}$ a vector of unknown

(Code 18.1)

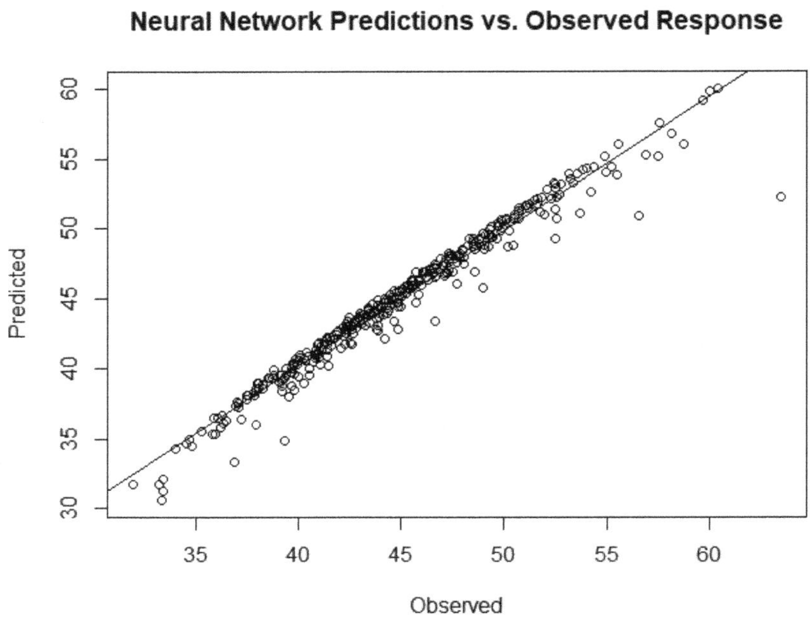

Fig. 18.2 In-sample (training data) predictions versus observed

"weights." The idea is that the input and response data are used to estimate the weights to minimize the error in associating the known inputs with their corresponding response values. In fact, a numerical least squares approach is often used to estimate the weights.

An example with nine regressors and $n = 400$ units was used to fit a neural network. Actually, the first 350 units were used to fit the network, and 50 units were saved to test the model.

Figure 18.2 shows the neural network "in-sample" predictions plotted against the observed responses, for the "training" set of 350 observations.

The "test" data, i.e., the last 50 samples, were used to first compute predictions from the neural network, and then plotted against the observed response value. Figure 18.3 shows the plot of the predicted and observed response variables for the out-of-sample or test data set.

In both Figs. 18.2 and 18.3, the line is the regression of predicted against observed.

(Code 18.1)

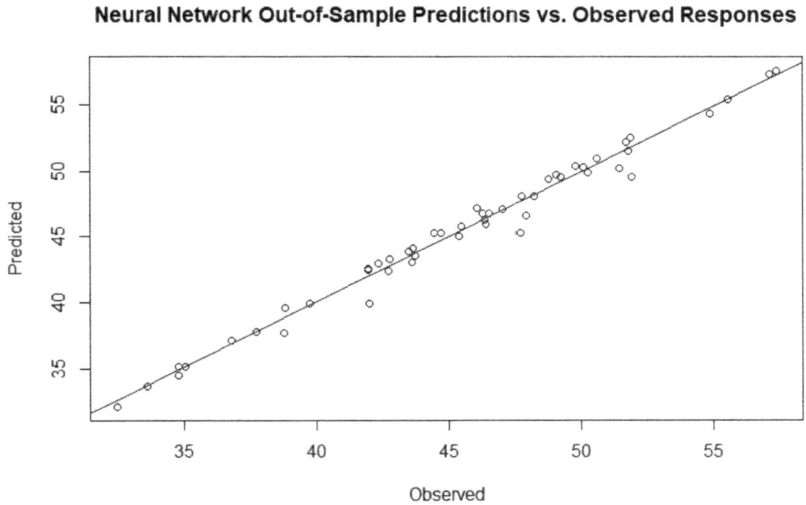

Fig. 18.3 Out-of-sample (test data) predictions versus observed

18.3 Classification and Regression Trees (CART)

CART is a way of generating predictions using linear functions (Breiman, et al., 1984). It resembles a "tree" since it has stages (like stepwise regression) and at each stage the data are split into two branches. The split is based on a regressor variable that produces two models with all the remaining regressors in such a way as to minimize the root mean square error (RMSE). The splitting continues until the sample size left is "too small" for splitting. CART is particularly useful when there are many regressors, so that an ordinary multiple linear regression model would be highly over-parameterized. CART creates a "tree," where each branch will lead to a predicted value.

Table 18.1 provides the progression of steps taken by CART to obtain a final model.

This CART algorithm used C_p as a criterion for splitting. There were a total of ten splits, and CART "decided" that regressors X0, X3, and X9 were the only ones worth using to make splits.

Figure 18.4 illustrates a regression tree from CART.

Figure 18.5 shows the predicted response plotted against the observed values, for the "training" set of the first 350 observations.

Figure 18.6 shows the CART predicted response values plotted against the observed values in the "test" set of the last 50 observations, which were not used in building the CART model.

Figure 18.7 shows the CART and neural network predictions for the test data set.

Table 18.1 CART steps and splits

(Code 18.2)

Regression tree:
rpart(formula = Response ~ X0 + X1 + X2 + X3 + X4 + X5 + X6 + X7 + X8 + X9, data = train.set)

Variables actually used in tree construction:
[1] X0 X3 X9

Root node error: 10350/350 = 29.571

n= 350

Step	C_p	nsplit	rel error	xerror	xstd
1	0.480327	0	1	1.00679	0.077549
2	0.088284	1	0.51967	0.56477	0.066344
3	0.085998	2	0.43139	0.51688	0.064622
4	0.066663	3	0.34539	0.44879	0.059803
5	0.0483	4	0.27873	0.38111	0.053301
6	0.039173	5	0.23043	0.32386	0.048987
7	0.022125	6	0.19126	0.24989	0.03618
8	0.013278	7	0.16913	0.22276	0.035894
9	0.011685	8	0.15585	0.1977	0.024913
10	0.01	10	0.13248	0.19598	0.024875

(Code 18.2)

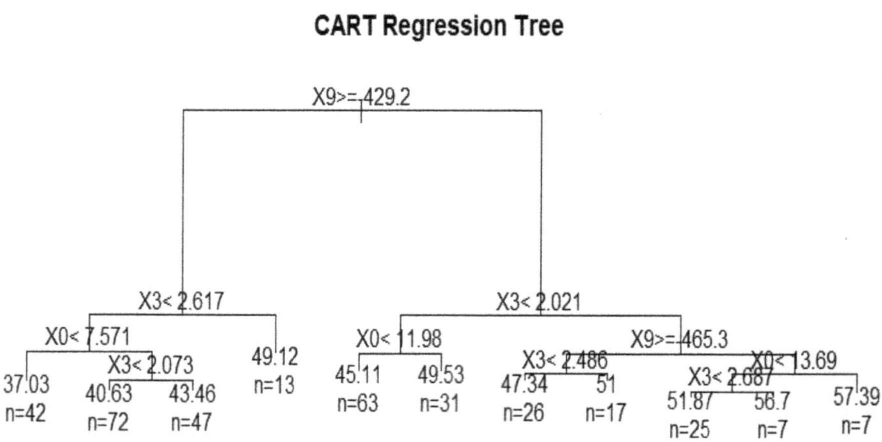

Fig. 18.4 CART regression tree

(Code 18.2)

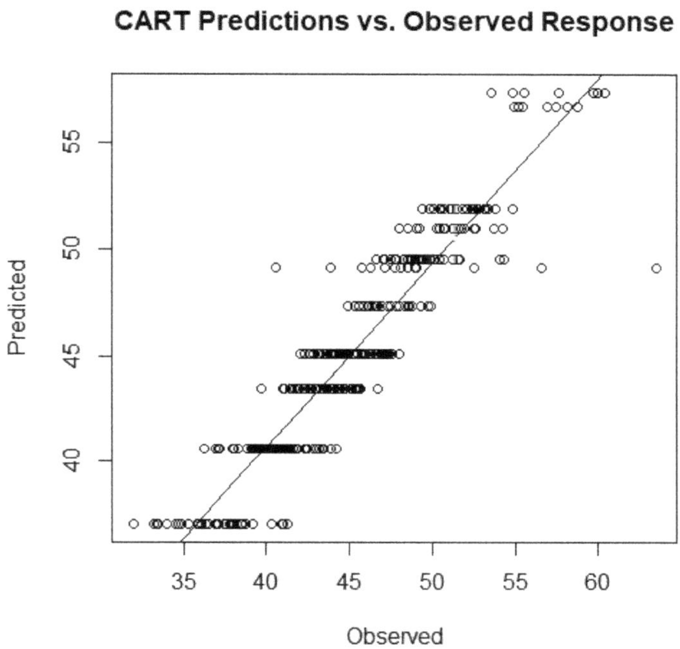

Fig. 18.5 CART predicted response vs. observed response—training data

(Code 18.2)

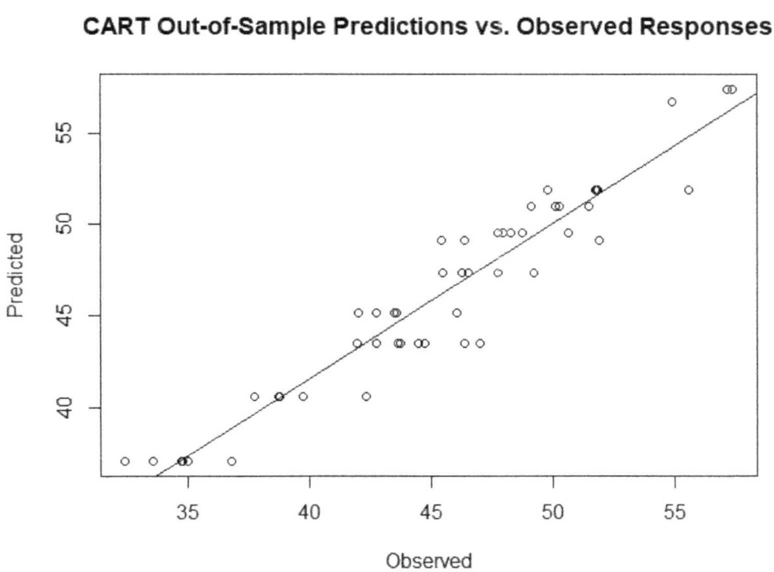

Fig. 18.6 CART predicted response vs. observed response—out-of-sample (test) data

(Code 18.3)

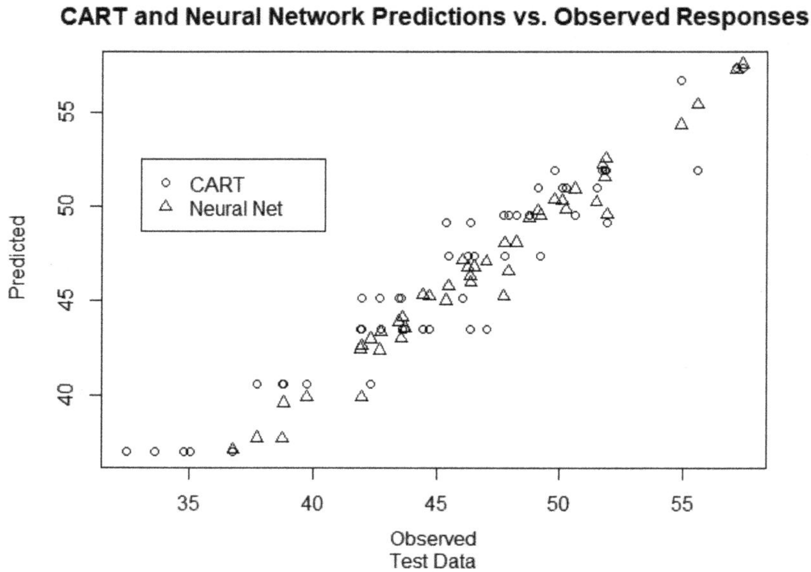

Fig. 18.7 CART and neural network predictions vs. observed response—test data

18.4 Random Forests

A Random Forest is a bootstrapping of CART) As a result, it is better than CART, inasmuch as it is many CART trees "averaged." There are many options in fitting a Random Forest model, including how many regressors to include on any of thew bootstrap samples and how many trees to construct. The interest reader might consult Breiman, 2001. There are many other works, some less mathematical than others. Figure 18.8 shows the out-of-sample predictions for the CART, network, and random forest models, which were all "trained," i.e., fit, with the same n = 350 observations. The predictions are for the 50 observations not included in the "training" set.

A feature of random forest models is the notion of importance for each regressor. The measure of importance is based on the number of bootstrap samples in which the regressor was included. The higher the value, the more important the regressor was in generating predictions. Table 18.2 provides the importance for the nine regressors in the random forest fit to the n = 350 observations in the training data set.

Recall that CART found regressors X0, X3, and X9 to be the only ones used for splitting.

(Code 18.3)

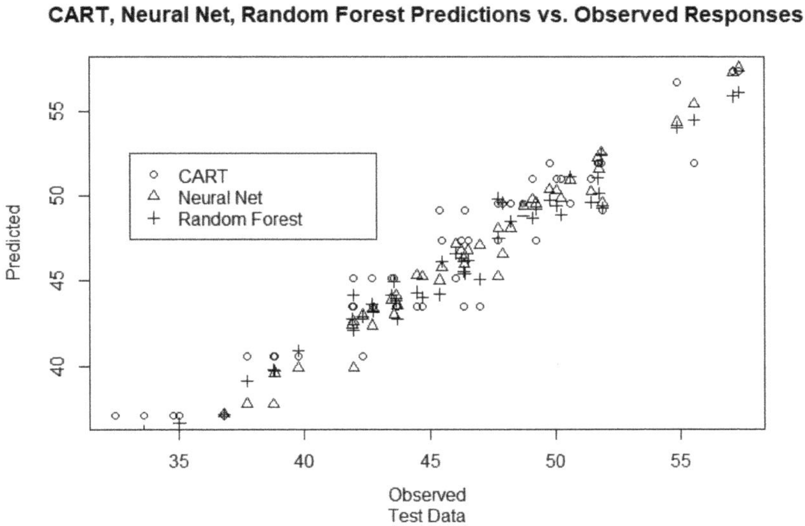

Fig. 18.8 CART, neural net, random forest predictions vs. observed—test data

Table 18.2 Random forest importance measures

Regressor	Importance
X9	**2723.5217**
X7	**2318.169**
X0	**2308.8832**
X3	**1868.1056**
X8	180.8905
X1	174.4571
X2	164.8495
X5	130.8444
X6	129.2378
X4	125.4894

18.5 Summary

Machine learning methods make use of data to fit models, where the model is chosen by an algorithm. Some methods described earlier, such as stepwise regression, Bayesian model averaging, linear discriminant analysis, and cluster analysis, could be considered machine learning methods. Perhaps the greatest utility of these methods is when there are very many explanatory/regressor variables, and there is no knowledge underlying their relationships to the response, or to the groups or categories in which individual experimental units might belong. As such, the machine learning methods do not necessarily provide the experimenter with any insight as to how the regressors are related to the response. The random forest method does provide a ranking of the importance of the regressors, which could then lead to designed experiments and greater insights.

18.6 Points to Ponder

1. Do machines learn?
2. Are there any drawbacks to using data-intensive "machine learning" methods?
3. What are some advantages of machine learning methods?

References

Breiman, L. (2001). Random forests. *Machine Learning, 45*, 5–32.

Breiman, L., Friedman, J., Stone, C. J., & Olshen, R. A. (1984). *Classification and regression trees*. Chapman and Hall/CRC.

Kuhn, M., & Johnson, K. (2013). *Applied predictive modeling*. Springer.

Chapter 19
Time Series and Dynamic Systems

19.1 Introduction

Most of the discussions in this text have been assuming that data form a random sample. That is, the set of observations $y_1, y_2, \ldots, y_n$ are all independent of each other, and they all have come from the same population. Sometimes, when data are gathered over time, the *i*th measurement, y_i, is correlated with the previous values. Such correlation is referred to as "autocorrelation" and requires special considerations.

19.2 Time Series

Time series will be defined here as a sequence of continuously valued random variables. For this text, we will assume that the variables are all normally distributed. The interesting part is that the values in the sequence are assumed to have non-zero correlations with each other. That is, the value observed at time t will have some correlation with previous values. These correlations will be exploited to help identify the underlying relationship that the values have with each other over time. Since the sequence is generally a set of values of the same measurement, but at different times, the correlations will be referred to as autocorrelations.

Suppose we observe a variable at uniformly-spaced time intervals, for n intervals, and represent these values as:

$$y_1, y_2, y_3, \ldots, y_t, y_{t+1}, \ldots, y_n$$

Supplementary Information The online version contains supplementary material available at https://doi.org/10.1007/978-3-031-84479-9_19. The videos can be accessed individually by clicking the DOI link in the accompanying figure caption or by scanning this link with the SN More Media App.

S. A. Pardo, *Beyond the T-Test*, https://doi.org/10.1007/978-3-031-84479-9_19

Mathematical statisticians would think of this as a realization of a sequence of random variables (Grimmett & Stirzaker, 2004). That is, suppose Y_1 was the first variable in the sequence. The value observed for Y_1 would be y_1. However, before Y_1 is observed, there are an infinite number of possible values it might have. Once Y_1 is observed, we think of it as being "realized." Suppose we further suspect that the value of y at time t is some unknown function not of time itself, but of the previous values of y, such as:

$$y_t = \beta_0 + \beta_1 y_{t-1} + \beta_2 y_{t-2} + a_t$$

The β_i are (unknown) constants, and a_t is a random "noise" variable with a mean value of 0 and a standard deviation, σ, that does not change with time. Furthermore, the a_t are assumed to be independent of each other as well as identically distributed. We will almost always assume that the noise variable is normally distributed, but the fixed mean and standard deviation assumptions are the most critical. It is also possible that the value of y at the "current" time, t, is expressed as a function of noise values at previous times:

$$y_t = \alpha_0 + \alpha_1 a_{t-1} + \alpha_2 a_{t-2} + a_t$$

Again, a_t is a random "noise" variable, and the α_i are constants. Think of the values a_{t-1} and a_{t-2} as random variables whose values have (somehow) already been observed and recorded. The values of a_t and y_t have not yet been observed. We will restrict the discussion to those time series that are a linear function of previous values and noise values.

Both of the expressions described above are what we will call "second order," since they only reach two steps back in time. The increment for terms in previous times are referred to as "lags." So terms with the subscript $t - 1$ are referred to as "lag 1" terms, with subscript $t - 2$ are referred to as "lag 2" terms, etc. The first expression is referred to as an autoregressive model (symbolized as AR); the second is referred to as a moving average (symbolized as MA) model (Cryer, 1986). It is also possible that the future values of the series are related to both the series past values and the past values of the random "noise":

$$y_t = \delta + \beta_1 y_{t-1} + \beta_2 y_{t-2} + \alpha_{t-1} a_{t-1} + \alpha_{t-2} a_{t-2} + a_t$$

Such a "hybrid" expression, or model, is called an autoregressive moving average (ARMA). The order of the models (how far back they reach in time) is designated with integers, e.g., AR(2) is a second order autoregressive model, MA(2) is a second order moving average model, and ARMA(2,2) is an ARMA model of orders 2 and 2.

19.3 Identifying Time Series Model Types and Orders

Box and Jenkins (Box, et al., 2008) discovered a methodology for identifying an appropriate model for time series. Their method involves computing two functions of the series data, called the autocorrelation function (*acf*) and partial autocorrelation function (*pacf*). First, some definitions are required. Recall the notion of expectation, or expected value. The covariance of two random variables, X and Y, is was defined earlier to be:

$$\text{cov}(X,Y) = E\left[(X-\mu_X)(Y-\mu_Y)\right] = \int_{-\infty}^{+\infty}\int_{-\infty}^{+\infty}(x-\mu_X)(y-\mu_Y)f(x,y)\,dxdy$$

The correlation between X and Y is a sort of standardized covariance:

$$\text{Corr}(X,Y) = \frac{\text{cov}(X,Y)}{\sigma_X^2\sigma_Y^2}$$

One could show that the correlation function is always a number between −1 and +1. In general, the correlation is positive if one variable increases when the other increases. It is negative is one variable decreases when the other increases. For time series, we are interested in how the values of the variable are related to values of the same variable, but observed at different times. So, for example, consider the values of a series at time t and at time $t-k$. The covariance would be:

$$\text{cov}(Y_t,Y_{t-k}) = E\left[\left(Y_t-\mu_{Y_t}\right)\left(Y_{t-k}-\mu_{Y_{t-k}}\right)\right]$$
$$= \int_{-\infty}^{+\infty}\int_{-\infty}^{+\infty}\left(y_t-\mu_{Y_t}\right)\left(y_{t-k}-\mu_{Y_{t-k}}\right)f(y_t,y_{t-k})\,dy_y dy_{t-k}$$

Since this covariance is using the same variable, just observed at different times, it is called the autocovariance. Accordingly, the autocorrelation is:

$$\text{Corr}(Y_t,Y_{t-k}) = \frac{\text{cov}(Y_t,Y_{t-k})}{\sigma_{Y_k}\sigma_{Y_{t-k}}}$$

In this text, we will only concern ourselves with time series that have a particularly convenient property; namely, that the autocorrelation only depends on the length of time between the time points, k, and not the specific time itself, t. Such time series will be referred to as "stationary". This is not the strictest, most general definition of stationarity, but it is the one on which we will rely. Stationarity is a necessary condition for being able to identify the particular form and order of model to be fit to the data. The subscript k is called the "lag". If $k = 0$, the autocorrelation is 1, and the covariance is the variance. For stationary time series, the covariance of lag k could be estimated as:

$$\widehat{acov}(k) = \frac{\sum_{t=k+1}^{n}(y_t-\bar{y})(y_{t-k}-\bar{y})}{n}$$

This expression is only a function of the lag, k. That is, the assumption of stationarity is required for this estimator to be meaningful. Also, the average, $\bar{y}$, is taken over all the observations, y_t, $t = 1, n$. Recall that this estimator is only a reasonable estimator if the time series is stationary.

The autocorrelation function (*acf*) can be estimated as:

$$r(k) = \widehat{acf}(k) = \frac{\widehat{acov}(k)}{\hat{\sigma}^2}$$

where:

$$\hat{\sigma}^2 = \frac{\sum_{t=1}^{n}(y_t - \bar{y})^2}{n}$$

A function of the lag that is closely related to the *acf* is called the partial autocorrelation function (*pacf*), mentioned earlier. The *pacf* is a sort of conditional *acf*, and in general is the conditional expectation (Cryer, 1986):

$$\phi(k) = pacf(k) = acf(k | y_{t-1}, y_{t-2}, \ldots, y_{t-k+1})$$

The estimated *pacf* values are obtained from the sample estimates of the *acf*, via a system of equations called the Yule-Walker equations (Box, et al., 2008):

$$r(j) = \varphi_1 r(j-1) + \varphi_2 r(j-2) + \ldots + \phi(k) r(0), \quad j = 1, k$$

That is, the coefficient for the 0th value of the *acf* is the estimate of the *pacf* at lag *k*. Of course, the 0th value of the *acf* is always 1. The unknown coefficients, φ_i, are actually the coefficients for an autoregressive model. The important point is that the *acf* and *pacf* are functions of the lag for stationary time series. The estimates of these functions will be used to identify a model to fit to the data.

19.4 The Box–Jenkins Approach

Box and Jenkins discovered that different forms of ARMA times series had specific morphologies for their associated *acf* and *pacf* functions. Thus, if one can estimate the *acf* and *pacf* functions of a time series, one can then determine the appropriate time series model to fit to the data. Moving average models of order *k* (MA(*k*)) models have *acf* functions that "cut off" (i.e., drop to zero) after lag *k*, and *pacf* functions that "taper off" toward zero as the lag gets larger and larger. Autoregressive time series of order *k* (AR(*k*)) will have *acf* functions that taper off to zero as the lag gets large, and *pacf* functions that will drop to zero after lag *k*. An autoregressive, moving average model of orders *k* and *l* (ARMA(*k*,*l*)) will have a *pacf* function that drops to zero after lag *k* and an *acf* function that drops to zero after lag *l*.

To illustrate the *acf* and *pacf* of various time series, we simulated three different time series. The first is called "white noise"; it is a non-autocorrelated sequence of random variables all having exactly the same distribution, generally with mean 0 and some standard deviation. If the variables are normally distributed, as in the case simulated here, the sequence is referred to as Gaussian white noise. In the case of white noise, we would expect the values of the *acf* and *pacf* to be patternless and relatively small. The second series is an AR(2). It is expected to have a significant partial autocorrelation at lags 1 and2, and then have only small values below the 95% confidence limits. It is expected to have an *acf* that looks like a damped sine wave; that is, it starts large but gradually oscillates toward 0. The third series is an MA(2). It is expected to have a significant autocorrelation at lags 1 and 2, and then the autocorrelations "cut off" and become non-significant. Its *pacf* is expected to oscillate like the *acf* for the AR(2) series.

The models are shown here:

White Noise:

$$y_t = \beta_0 + a_t$$

AR(2):

$$y_t = \beta_0 + \beta_1 y_{t-1} + \beta_2 y_{t-2} + a_t$$

MA(2):

$$y_t = \alpha_0 + \alpha_1 a_{t-1} + \alpha_{t-2} a_{t-2} + a_t$$

The terms in the models with the subscript 0 are called "intercepts."

Table 19.1 provides the parameter estimates for the three series.

Figures 19.1 and 19.2 show the *acf* and *pacf* of the white noise series. Figures 19.3 and 19.4 show the *acf* and *pacf* for the AR(2) series. Figures 19.5 and 19.6 show the *acf* and *pacf* of the MA(2) series. The dashed horizontal lines on the *acf* and *pacf* plots are approximate 95% confidence limits for the correlations. Any correlation within the two lines is not significantly difference from 0.

These time series models look a lot like multiple linear regression models, with regressors that had a time component to them. That is because they are in fact multiple linear regression models. The estimation is done slightly differently than OLS regression, to account for autocorrelation, but the principles are the same.

Table 19.1 Parameter estimates for time series models

Series	Intercept	Lag 1	Lag 2
White Noise	-0.0224	0.0000	0.0000
AR2	-0.0840	0.7346	-0.5274
MA2	0.0144	0.8506	0.0648

Fig. 19.1 ACF—white noise series

(Code 19.1)

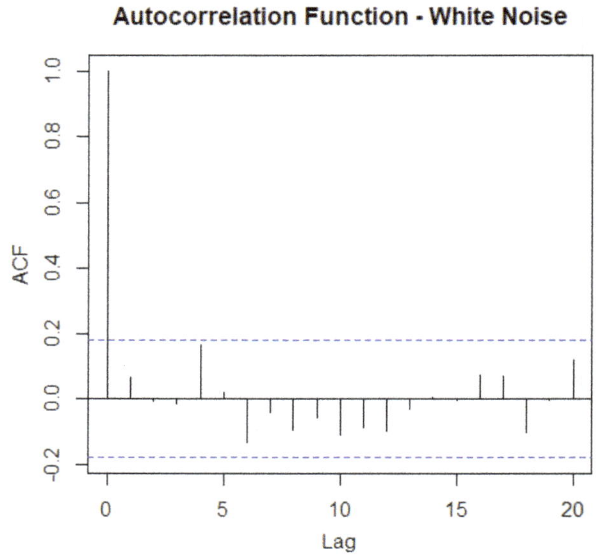

Fig. 19.2 PACF—white noise series

(Code 19.1)

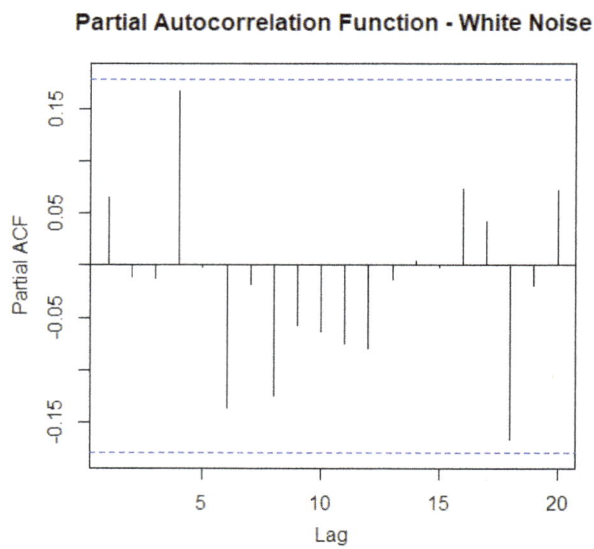

Fig. 19.3 ACF—AR(2) series

(Code 19.1)

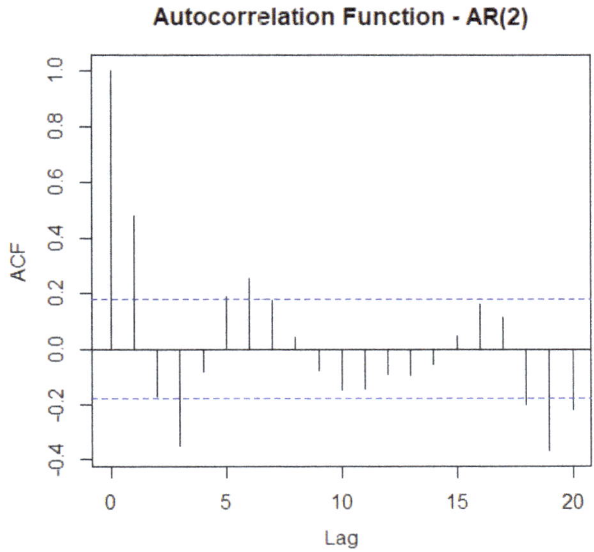

Fig. 19.4 PACF—AR(2) series

(Code 19.1)

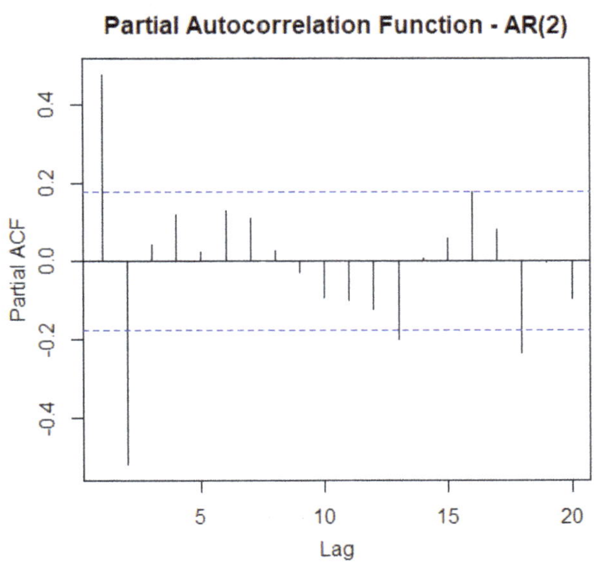

Fig. 19.5 ACF—MA(2) series

(Code 19.1)

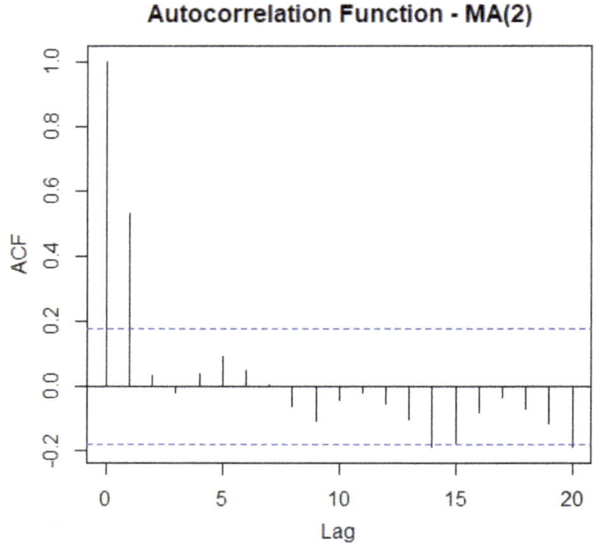

(Code 19.1)

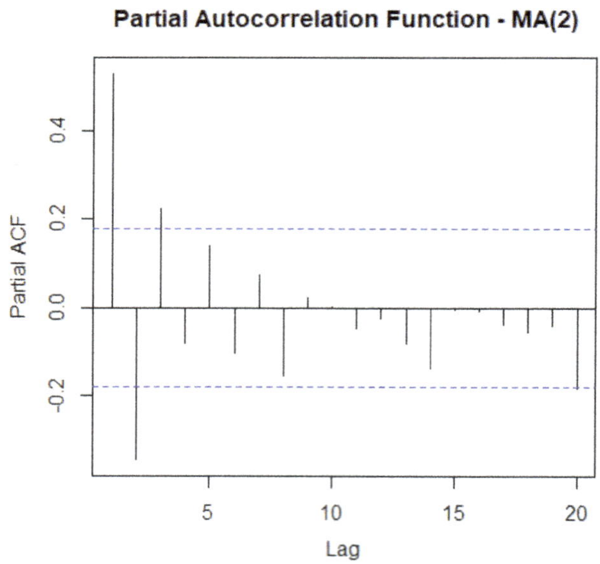

Fig. 19.6 PACF—MA(2) series

19.5 Non-stationarity and Differencing

One of the primary assumptions required for the Box–Jenkins model identification process is stationarity. For our purposes, stationarity refers to the phenomenon where the correlation in the series depends only on the separation in time, or lag, and not on the specific times. There is, however, a class of non-stationary time series where stationarity can be induced via a simple transformation called differencing. If Y_t is a time series, then define the difference of order 1 to be:

$$D_t = Y_t - Y_{t-1}$$

A difference of order 2 would be a difference of differences:

$$D_t^{(2)} = D_t - D_{t-1}$$

In general, an order d difference would be:

$$D_t^{(d)} = D_t^{(d-1)} - D_{t-1}^{(d-1)}$$

There are situations where such differenced series are stationary although the original series are not. Consider, for example, the following series:

$$y_t = y_{t-1} + \beta_2 y_{t-2} + a_t$$

This series is in fact non-stationary. However, the series:

$$d_t = y_t - y_{t-1} = \beta_2 y_{t-2} + a_t$$

may in fact be stationary, depending on the value of β_2.

Typically, non-stationary time series, before any differencing, have very large *acf* and *pacf* values. Figure 19.7 shows an *acf* for a random walk.

An ARMA model that is based on *k*th order differences is called an ARIMA model, where the "I" stands for "integrated." The "integration" is how the predicted values of the original series are obtained from the model fit to the differences. The notation to express the order of an ARIMA model is ARIMA(k,d,l) where, as in the case of ARMA models, k is the order for the autoregressive part, l is the order for the moving average part, and d is the order of differencing.

A special case of a non-stationary process is the "random walk" (Cryer, 1986). The random walk process can be described as:

$$y_t = y_{t-1} + a_t$$

The first-order difference is:

$$d_t = y_t - y_{t-1} = a_t$$

(Code 19.1)

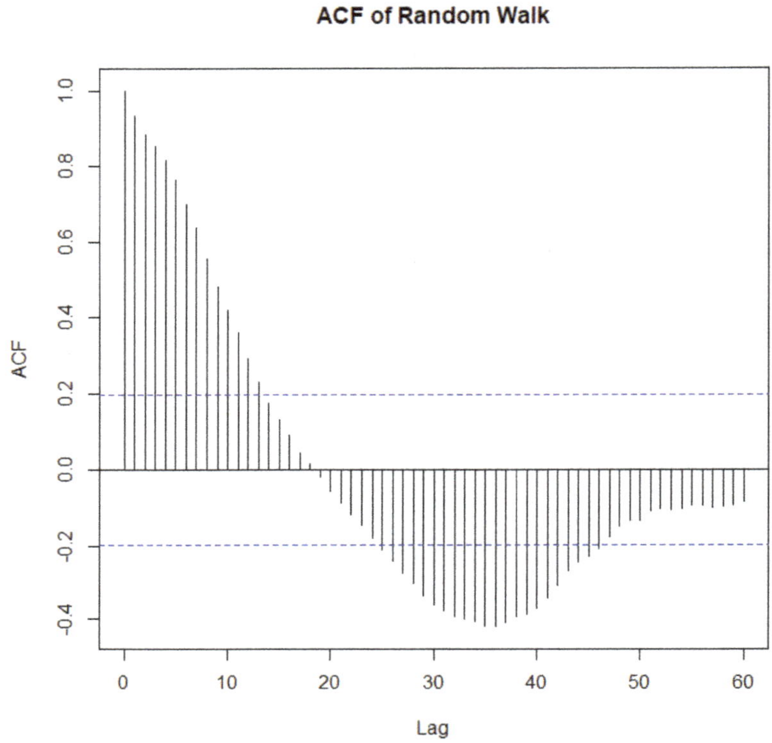

Fig. 19.7 ACF for a random walk process

In other words, differencing yields a process that is simply noise. This is why random walks are inherently unpredictable. In fact, the best prediction or forecast for the process value at time t is whatever its value was at time $t - 1$. The fact that the value at time t only depends on the immediately preceding time, $t - 1$, makes the random walk something called a Markov process (Grimmett & Stirzaker, 2004),

Often time series models are used to "forecast" beyond the last time at which data were observed. The use of such forecasts can be helpful, especially if the underlying data-generating process is really well-modeled by the ARIMA model chosen. However, be aware that certain models are inherently unpredictable, especially "random walk" models. As mentioned earlier, the best forecast for a random walk is the immediately preceding value of the series. So the best forecast for y_t is y_{t-1}. Stock prices commonly follow a random walk model.

19.6 A Note on Frequency Domain Analyses

Prior to the conception of the Box–Jenkins methodology, time series were largely analyzed using Fourier methods (Bloomfield, 2000). The basic idea is that the series was modeled as a sum of terms involving sines and cosines:

$$y_t = \alpha_0 + \sum_{k=1}^{T} \beta_k \sin(\omega_k t) + \sum_{k=1}^{T} \gamma_k \cos(\omega_k t) + a_t$$

and

$$\omega_k = \frac{2\pi k}{T}$$

While this sort of analysis has many useful applications, it will not be discussed here. However, there is a section in the appendix on mathematical concepts with more of a discussion about "frequency domain" analyses (the ω_k are called frequencies). For those who are interested seem Bloomfield (2000).

19.7 Markov Chains

A Markov Chain is a sequence of discrete random variables, $X_1, X_2, \ldots$, changing in discrete time steps, that has a special property:

$$\Pr\{X_k = x_k | X_0 = x_0, X_1 = x_1, \ldots X_{k-1} = x_{k-1}\} = \Pr\{X_k = x_k | X_{k-1} = x_{k-1}\}$$

In other words, the probability that the current value in the sequence has any particular value only depends on the immediately preceding value, and not on the entire "history" of the sequence. This property is called the "Markov Property" (Grimmett and Stirzaker, 2004).

As mentioned earlier, the random walk is a Markov process, which is more general than a Markov chain, except the random variables may be continuously valued.

For the sake of simplifying notation, we will refer to the values, or "states" of the X_k as 1, 2, 3, …, e.g., $X_k = 2$, although there is not necessarily any order to the states. For example, a traffic light can have the "states" of red, yellow, or green. We could arbitrarily assign to the "red" state the value 1, "yellow" could be 2, and "green" could be 3. For further simplification, we will use the notation for the conditional probabilities:

$$\Pr\{X_k = j | X_{k-1} = i\} = p_j$$

For processes having the Markov Property, p_{ij} is the conditional probability that the next state will be j given that the current state is i. These conditional probabilities are called transition probabilities. Most of the chains we will discuss have only a finite number of possible states, as in the case of traffic lights. Thus, the one-step transition probabilities for an n-state chain can be represented by a (finite-dimensional) n x n matrix:

$$\boldsymbol{P} = \begin{bmatrix} p_{11} & p_{12} & \cdots & p_{1n} \\ \vdots & \ddots & \ddots & \vdots \\ p_{n1} & p_{n2} & \cdots & p_{nn} \end{bmatrix}$$

As we have defined the transition probability matrix, each row represents a conditional distribution for the states $j = 1, n$, i.e.,

$$\sum_{j=1}^{n} p_{ij} = 1$$

Keep in mind that it is thoroughly possible that there are some states such that:

1. $p_{ij} = 1$
2. $p_{ij} = 0$
3. $p_{ij} = 0$ for $j \neq i$ and $p_{ii} = 1$

The condition in (3) is referred to as "absorbing," so that state i is called an "absorbing state." That is, once the chain reaches state i, it can never transition to any other state.

There are several interesting results and consequences related to finite-state Markov Chains. It all stems from the Markov property. The matrix $\boldsymbol{P}$ is called the one-step transition probability matrix. The transition probabilities for 2, 3, 4, etc. steps ahead can be calculated as a power of matrix $\boldsymbol{P}$. If $p_{ij}(1)$ represents the transition probability from state i to state j in one time step, and $\boldsymbol{P}(k)$ is the matrix of k-step transition probabilities, then:

$$\boldsymbol{P}(k) = \boldsymbol{P}^k$$

In many cases, as k gets very large, there is a convergence so that each row is identical, and the row represents the long-run probability of finding the "system" in each of the states. In order to have such convergence, the chain must be "irreducible," meaning that there are no absorbing states. In a gambling game, for example, once the gambler reaches the state of no money, the gambler can no longer play. Thus the state of no money is an absorbing state. The long-run probability distribution is called the steady state, or sometimes "stationary," distribution, and can be calculated. From this steady-state distribution, the expected, or average, state can also be calculated.

Table 19.2 A 1-Step transition matrix: 4-state markov chain

	S1	S2	S3	S4
S1	0.879	0.060	0.060	0.001
S2	0.000	0.110	0.840	0.050
S3	0.000	0.840	0.110	0.050
S4	0.001	0.000	0.000	0.999

Table 19.3 Stationary distribution for 4-State Markov Chain

(Code 19.3)

States:	S1	S2	S3	S4
Steady-State Probabilities	0.00804	0.00965	0.00965	0.97267

19.8 Steady State, or "Stationary" Distribution

To compute the stationary distribution, the 1-step transition matrix must be known (or at least estimated from data). Table 19.2 provides the 1-step transition matrix for a 4-state Markov Chain.

The stationary distribution for this Markov Chain is given in Table 19.3.

So we could expect this chain to be is state **S4** most (>97%) of the time.

19.9 Summary

Dynamic models are those whose values are expected to change over time. When those values are dependent on previous values, the data-generating process is referred to as autocorrelated. A special family of autocorrelated processes are linear time series, which are commonly referred to simply as time series. There is a family of linear time series called AR. They are very flexible and can be applied in a very wide variety of applications. The Box and Jenkins approach helps to identify the particular form of ARIMA model to fit to data.

When the values in a sequence are discrete, and in fact only a finite number of possible values or "states," of interest is the probability of "transitioning" from one state to each other state. When the probability of going to a state only depends on the immediately preceding state, the process is called "Markov." A process of transition from one discrete state to another, when it has the Markov property, is called a Markov Chain. One feature of Markov chains is that they may have a "stationary" or steady-state probability of being in each state. Such a stationary distribution can be derived from the matrix of transition probabilities.

19.10 Points to Ponder

1. What is autocorrelation?
2. How are autocorrelation and partial autocorrelation used to determine models for linear time series?
3. What is the Markov property?
4. Can you name a Markov Chain you have observed?
5. Why would a stationary distribution be useful?

References

Bloomfield, P. (2000). *Fourier Analysis of Time Series: an Introduction* (2nd ed.). Wiley.

Box, G. E. P., Jenkins, G. M., & Reinsel, G. C. (2008). *Time Series Analysis: Forecasting and Control* (4th ed.). Wiley.

Cryer, J. D. (1986). *Time Series Analysis*. Duxbury Press.

Grimmett, G., & Stirzaker, D. (2004). *Introduction to Probability and Random Processes* (3rd ed.). Oxford University Press.

Appendices

1.1 Appendix A: Mathematical Concepts

1.1.1 A1. Basics of Calculus

1.1.1.0.1 Limits

A limit is a sort of boundary to some sort of numerical progression. Clearly, some progressions are infinite, i.e., they have no bound. Their limit is said to be infinity ($+\infty$), or negative infinity, ($-\infty$). An example of a progression is:

$$f\left(k\right)=\frac{1}{k},k=1,2,3,\ldots$$

As k gets large (or, as we say, goes to $+\infty$), $f(k)$ gets smaller and smaller. Is it ever 0? No, but it gets closer and closer to 0 as k goes to $+\infty$. Thus, the limit of $f(k)$ is said to be 0.

Here is another example. The progression (or, as mathematicians like to call it, sequence):

$$f\left(k\right)=k,k=1,2,3,\ldots$$

Just keeps getting larger as k goes to $+\infty$. Thus, its limit is $+\infty$. When a sequence has as its limit either $\pm\infty$, it is said to have no limit, or its limit does not exist, or it is unbounded.

The above examples were sequences of positive integers but you could have sequences of real numbers, too. It is just that you can't write the sequence of the "input" values by separating numbers with commas. Why? Because between any two real numbers, no matter how close they are to each other, there is always another

S. A. Pardo, *Beyond the T-Test*, https://doi.org/10.1007/978-3-031-84479-9

real number in between. So, to represent the sequence of input values when they are real numbers, we use inequality signs. For example:

$$f(x) = \frac{1}{x}, x > 0$$

Guess what is the limit of the sequence $f(x)$? Did you say 0? Give the player a cupie doll!

The two important ideas of calculus, derivatives, and integrals are based on limits of real-valued sequences.

1.1.1.1 A2. Derivatives and Differentiation

Take a nice, smooth function of real numbers, call it $f(x)$. Suppose you pick and value of x, and then another value a little to the right of x, call it $x + h$ (h is some fixed amount). The difference in the function from x to $x + h$ is:

$$d(x,h) = f(x+h) - f(x)$$

So d is the amount the function changed when moving from x to $x + h$. The thing is, if h is very big, you might get a different amount of change in f than if h is very small. So we like to talk about the change in f per unit of x. This could be expressed as:

$$\delta(x,h) = \frac{f(x+h) - f(x)}{x+h-x} = \frac{f(x+h) - f(x)}{h}$$

Note that h is in the same units as x.

The function $\delta(x,h)$ might have a limit as h is made smaller and smaller (closer and closer to 0, but not 0 itself) This limit is called the derivative of f at x. It turns out that many common functions have formulas that describe these limits, or derivatives. The derivative of f at x tells you the "slope" of the function, or how fast it is changing when the input value is "near" x. In one case, the derivative is the same number no matter what the value of x; this special class of functions can be described as:

$$f(x) = a + bx, \text{ where } a \text{ and } b \text{ are two constants.}$$

That is correct; this is a straight line function with intercept a and slope b. The slope, b, is the value of the derivative no matter what the value of x is.

There are several notations for expressing the derivative of a function. If the function is $f(x)$, then its derivative (which is in turn a function itself) is expressed as:

$$\frac{df(x)}{dx}$$

Or

$$f'(x)$$

Or

$$\dot{f}(x)$$

Those are the three most common notations.

It turns out that certain common functions have formulas for their derivatives. Here are a few:

$f(x)$	$\frac{df}{dx}$
ax^n	anx^{n-1}
$\frac{1}{x^n}$	$\frac{-n}{x^{n+1}}$
$\sin(x)$	$\cos(x)$
$\cos(x)$	$-\sin(x)$
e^{ax}	ae^{ax}
$\log_e(x)$	$\frac{1}{x}$

For a special case of ax^n, consider:

$$f(x) = ax$$

Then:

$$\frac{d}{dx}(ax) = ax^{1-1} = ax^0 = a*1 = a$$

The most important rules about derivatives are linearity and the chain rule:

$$\frac{d}{dx}\left(af(x) + bg(x)\right) = a\frac{df}{dx} + b\frac{dg}{dx}$$

And

$$\frac{d}{dx}f(g(x)) = f'(g(x))g'(x)$$

Notice how I changed notations from describing the linearity rule to the chain rule. It was just more convenient.

Example of linearity:

$$\frac{d}{dx}\left(ax^2 + bx^3\right) = 2ax^{2-1} + 3bx^{3-1} = 2ax + 3bx^2$$

Example of Chain Rule:

$$\frac{d}{dx}\sin\left(ax^n\right) = \cos\left(ax^n\right) * anx^{n-1}$$

A special case of linearity:

$$\frac{d}{dx}\left(af(x)\right) = a\frac{df}{dx}$$

Also, if a is a constant, and:

$$f(x) = a$$

regardless of the value of x, then:

$$\frac{d}{dx}\left(f(x)\right) = \frac{d}{dx}(a) = 0$$

Note that the derivative of a function is itself a function. As a result, its value can be computed for a given value of x. So, for example, suppose:

$$f(x) = 3x^2$$

So its derivative would be:

$$f'(x) = 2 * 3x^{2-1} = 6x$$

Thus, the value of the derivative, $f'(x)$, at $x = 1$ is:

$$f'(1) = 6(1) = 6$$

When a function has more than one "x" variable, then it has derivatives for each of the x's, independent of the other x's. So, suppose f is a function of two variables, call them x_1 and x_2.

Then there is a derivative of f with respect to x_1 and another derivative of f with respect to x_2. The notation commonly used is:

$$\frac{\partial f(x_1, x_2)}{\partial x_1}$$

and

$$\frac{\partial f(x_1, x_2)}{\partial x_2}$$

When forming these "partial derivatives," the variables not being included in the differentiation process (the process of forming the derivative) are treated like they were constants. A special but important case would arise in multiple linear regression. Suppose the function was:

$$f(x_1, x_2) = \beta_1 x_1 + \beta_2 x_2$$

Then:

$$\frac{\partial f(x_1, x_2)}{\partial x_1} = \beta_1$$

and

$$\frac{\partial f(x_1, x_2)}{\partial x_2} = \beta_2$$

This is just included so that you will be aware of the notation and have some familiarity with it.

1.1.1.1.1 Higher-Order Derivatives

If there is a function, $f(x)$ that has a derivative, then the derivative is also a function in the variable x. Thus, the derivative of f can also have a derivative. This derivative of the derivative of f is called the second derivative of f. This second derivative in turn can have a derivative, which is now the third derivative of f. This can go on forever. Keep in mind that the derivative of a function that is equal to a constant (just a single number) is 0, regardless of the value of the constant. Of course, if the constant is 0, then the derivative is still 0.

Here is an example. Suppose:

$$f(x) = 2 + 3x^4$$

Its derivative is:

$$f'(x) = \frac{df}{dx} = 3 * 4x^{4-1} = 12x^3$$

This function in turn has a derivative:

$$f''(x) = \frac{d^2 f}{dx^2} = 3 * 12x^{3-1} = 36x^2$$

We use the superscript "2" to indicate that this is the second derivative of the original function, *f*.

We can continue taking derivatives of the derivatives:

$$f'''(x) = \frac{d^3 f}{dx^3} = 2 * 36x^{2-1} = 72x^1$$

$$f''''(x) = \frac{d^4 f}{dx^4} = 1 * 72x^{1-1} = 72x^0 = 72$$

$$f'''''(x) = \frac{d^5 f}{dx^5} = 0 * 72 = 0$$

Once the nth order derivative (in this case, 5th) is 0, all subsequent derivatives will be 0.

1.1.1.2 A3. Derivatives and Optima

A function of a single variable is at its maximum or minimum when its first derivative is equal to 0. That means if there is a value of the variable that makes the first derivative equal to 0, then that is the value of the variable where the original function is either maximum or minimum.

Without even graphing the original function, you can determine whether the point at which its first derivative is 0 is either a maximum or minimum. You do this by taking the derivative of the derivative (this is the second derivative). If at the value of the variable making the first derivative 0, the second derivative is negative, then the function is at its maximum at that value of the variable. If the second derivative is positive, then this is a minimum for the original function.

We will illustrate this with a funny function:

$$f(p,q) = pq^2$$

This is a function of two variables, but suppose $q = 1 - p$. Then the function becomes:

$$f(p) = p(1-p)^2 = p(p^2 - 2p + 1)$$
$$= p^3 - 2p^2 + p$$

The derivative of this function is:

$$\frac{df}{dp} = 3p^2 - 4p + 1$$

To find the minimum or maximum of the original function, set the derivative equal to 0 and solve:

$3p^2 - 4p + 1 = 0$
$(3p - 1)(p - 1) = 0.$

So either $p = 1/3$ or $p = 1$. First, find out of these two values of p make the second derivative negative or positive.

The second derivative is:

$$\frac{d^2 f}{dp^2} = 6p - 4$$

Using $p = 1/3$ yields $6(1/3) - 4 = -2$. With $p = 1$, $6(1) - 4 = 2$. At what value is the second derivative in between being negative or positive? When it equals 0:

$6p - 4 = 0$
$p = 4/6 = 2/3.$

So, if $p < 2/3$, the second derivative is negative, so that at $p = 1/3$, the original function has a "local maximum" (a bump). If $p > 2/3$, the second derivative is positive, so at $p = 1$, the original function has a "local minimum" (a trough). At $p = 2/3$, the original function is changing from going up to going down (an inflection point).

Look at the plot:

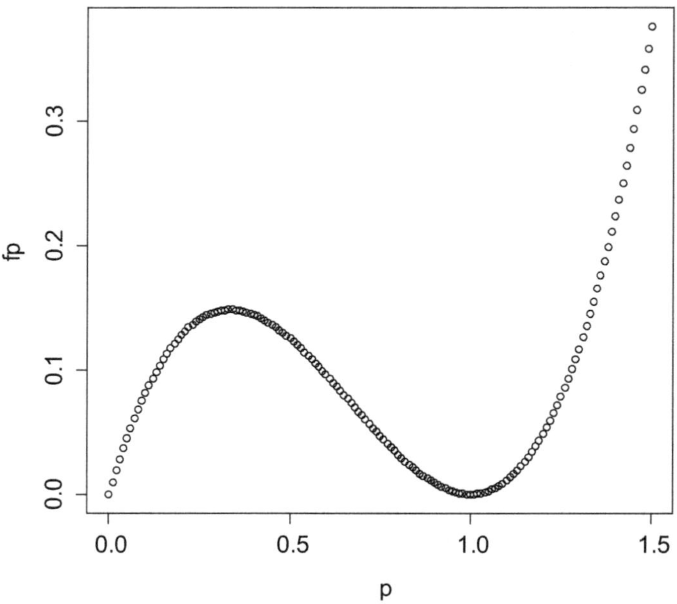

Note that the "trough" is at $p = 1$. If the variable p is restricted to the open "unit interval" (0, 1), then we may only be interested in the "local maximum" at $p = 1/3$. But the interval of interest depends on whether the function is a "model" describing some real phenomenon.

1.1.1.3 A4. Integrals and Integration

Suppose there was some function, $f(x)$, and you wanted to know how much area there was under it (assume for the moment that the function values are always non-negative) between two values on the x-axis (say x_1 and x_2, with $x_1 < x_2$). For now, don't ask why on earth you would want to know this area; just take for granted that there will be some good reasons. Suppose you broke up the range of values from x_1 to x_2 in the following fashion:

$x_1, x_1 + h, x_1 + 2\,h, x_1 + 3\,h, \ldots, x_2,$

The intervals between each consecutive value in this progression all have the same length, h. The function has values at each of the x's in the progression:

$f(x_1), f(x_1 + h), f(x_1 + 2\,h), \ldots, f(x_2),$

Suppose that we choose a value k such that $x_2 = x_1 + kh$. The area of a rectangle is width times height. Consider the rectangle starting at x1 with "base" width h, and with height $f(x_1)$. Its area would be $f(x_1)*h$. Then consider the next rectangle from $x_1 + h$ to $x_1 + 2\,h$, with height $f(x_1 + h)$. Its area would be $f(x_1 + h)*h$. Go through the whole progression, so that at any point, the height of the rectangle is $f(x_1 + kh)$ and its with is always h, so that its area is $f(x_1 + kh)*h$. Now add up the areas of all the little rectangles, and you have an approximate area under the function $f(x)$ between x_1 and x_2. Mathematically, this area can be expressed as:

$$L = \sum_{j=0}^{k-1} f(x_1 + jh)h$$

We could have also approximated the area a little differently. Instead of using the height of the first rectangle to be $f(x1)$, suppose we had started with $f(x_1 + h)$. The approximate area would then be:

$$U = \sum_{j=1}^{k} f(x_1 + jh)h$$

Suppose we try to find the limit of L as h gets smaller and smaller, and the limit of U as h gets smaller and smaller. If these two limits are not infinite, and they equal each other, then the limit is called the (Reimann) integral of $f(x)$ between x_1 and x_2. The usual notation for the integral of f between values x_1 and x_2 is:

$$A = \int_{x_1}^{x_2} f(\xi)d\xi$$

The Greek letter ξ acts like an index variable in summation notation. When the integral has a formula, call it $F(x)$, then the area, A, is computed as the difference $F(b) - F(a)$, which is sometimes denoted as:

$$F(b) - F(a) = F(x)\Big|_a^b$$

As in the case of derivatives, the integrals of some common functions have easy-to-compute forms. The table gives some examples:

$f(x)$	$\int_a^b f(x)dx$
anx^{n-1}	$ax^n\Big\|_a^b$
ae^{ax}	$e^{ax}\Big\|_a^b$
$\cos(x)$	$\sin(x)\Big\|_a^b$
$-\sin(x)$	$\cos(x)\Big\|_a^b$

$\frac{1}{x}$	$\log_e(x)\Big\|_a^b$
$\frac{-n}{x^{n+1}}$	$\frac{1}{x^n}\Big\|_a^b$

You may have noticed from this table and the table of derivative formulas that when a derivative has a formula, the integral is the original function that gave rise to the derivative. Sometimes the integral is referred to as the anti-derivative. This relationship is useful in solving a class of problems called differential equations. Differential equations are equations in which relationships between derivatives are known, but the original function is unknown. The integral is useful for finding the original function that satisfies the differential equation. As a simple case, suppose that we know:

$$\frac{df(x)}{dx} = \cos(x)$$

Then the original function must be:

$$\int \frac{df(x)}{dx} = \int \cos(x)\,dx = \sin(x)$$

We have not specified any limits on the integral, since we were just trying to illustrate the idea that the integral as anti-derivative is useful in solving differential equations. Such integrals are referred to as indefinite; often the resulting formula is written with the letter C as an added term:

$$\int \cos(x)\,dx = \sin(x) + C$$

The integration process has some very useful rules. Possibly the most useful is, like the derivative, linearity. Specifically:

$$\int af(x) + bg(x)\,dx = a\int f(x)\,dx + b\int g(x)\,dx$$

1.1.1.4 A5. Matrix and Vector Algebra

A matrix is a rectangular array of numbers:

$$\boldsymbol{M} = \begin{bmatrix} a_{11} & a_{12} & \dots & a_{1n} \\ \vdots & \vdots & \ddots & \vdots \\ a_{m1} & a_{m2} & \dots & a_{mn} \end{bmatrix}$$

The "entries," a_{ij}, can be real or complex numbers, but we will only deal with real matrices and vectors.

A vector is a special matrix that has either only one row or one column. For the most part, we will consider vectors to be a column of numbers:

$$\boldsymbol{v} = \begin{bmatrix} b_1 \\ \vdots \\ b_m \end{bmatrix}$$

We will represent the names of matrices with bold-face, capital letters, and the names of vectors with bold-face, lower case letters. We will use the convention of labeling the entries, or components, with subscripts that represent the row number (first subscript) and column number (second subscript). In the case of vectors, the single subscript will refer to the row number for "column" vectors or the column number for 'row" vectors. As we stated before, mostly when we refer to a vector, we will be thinking of it as a column. When we need to refer to a "row" vector, most of the time the name of the vector will have a superscript "T," which is an abbreviation for "transpose." A transpose vector or matrix is one where the rows become columns.

There are several operations involving matrices and vectors that will appear throughout the text. They are:

1. Scalar multiplication.
2. Matrix and vector addition.
3. Transposition.
4. Matrix multiplication.
5. "Dot" or scalar product of two vectors, and vector length
6. Square matrices, the identity matrix, and matrix inverses.
7. Determinants.
8. Eigenvalues and eigenvectors.
9. Diagonalization and powers of matrices.

1.1.1.5 A6. Scalar Multiplication

A scalar for our purposes is a number that is not a vector or matrix. In fact, the entries or components in all of our vectors and matrices will be scalars. To multiply a scalar by a matrix of vector, simply multiply that scalar by every entry in the matrix or vector. So:

$$\boldsymbol{A} = s\boldsymbol{M} = \boldsymbol{M}s = \begin{bmatrix} sa_{11} & sa_{12} & \dots & sa_{1n} \\ \vdots & \vdots & \ddots & \vdots \\ sa_{m1} & sa_{m2} & \dots & sa_{mn} \end{bmatrix}$$

or

$$\boldsymbol{w} = s\boldsymbol{v} = \boldsymbol{v}s = \begin{bmatrix} sb_1 \\ \vdots \\ sb_m \end{bmatrix}$$

The order is not important for scalar multiplication. This is not the case of matrix multiplication.

1.1.1.6 A7. Matrix and Vector Addition

Adding matrices and vectors is component-wise addition. This means that in order to add two matrices, they must have exactly the same number of rows and columns. So:

$$\boldsymbol{M} = \begin{bmatrix} a_{11} & a_{12} & \dots & a_{1n} \\ \vdots & \vdots & \ddots & \vdots \\ a_{m1} & a_{m2} & \dots & a_{mn} \end{bmatrix}$$

$$\boldsymbol{N} = \begin{bmatrix} b_{11} & b_{12} & \dots & b_{1n} \\ \vdots & \vdots & \ddots & \vdots \\ b_{m1} & b_{m2} & \dots & b_{mn} \end{bmatrix}$$

$$\boldsymbol{M} + \boldsymbol{N} = \boldsymbol{N} + \boldsymbol{M} = \begin{bmatrix} a_{11} + b_{11} & a_{12} + b_{12} & \dots & a_{1n} + b_{1n} \\ \vdots & \vdots & \ddots & \vdots \\ a_{m1} + b_{m1} & a_{m2} + b_{m2} & \dots & a_{mn} + b_{mn} \end{bmatrix}$$

and

$$\boldsymbol{v} = \begin{bmatrix} a_1 \\ \vdots \\ a_m \end{bmatrix}$$

$$\boldsymbol{w} = \begin{bmatrix} b_1 \\ \vdots \\ b_m \end{bmatrix}$$

$$\boldsymbol{v} + \boldsymbol{w} = \boldsymbol{w} + \boldsymbol{v} = \begin{bmatrix} a_1 + b_1 \\ \vdots \\ a_m + b_m \end{bmatrix}$$

1.1.1.7 A8. Transposition

If:

$$\boldsymbol{M} = \begin{bmatrix} a_{11} & a_{12} & \dots & a_{1n} \\ \vdots & \vdots & \ddots & \vdots \\ a_{m1} & a_{m2} & \dots & a_{mn} \end{bmatrix}$$

Then the transpose of matrix $\boldsymbol{M}$, call it $\boldsymbol{M}^T$, is:

$$\boldsymbol{M}^T = \begin{bmatrix} a_{11} & \dots & a_{m1} \\ a_{12} & \dots & a_{m2} \\ \vdots & \ddots & \vdots \\ a_{1n} & \dots & a_{mn} \end{bmatrix}$$

To illustrate the point, the subscripts on the entries in $\boldsymbol{M}^T$ have not been changed from the original, although our convention is to have the row number first followed by the column number. As a more concrete example, if:

$$\boldsymbol{M} = \begin{bmatrix} 1 & 2 & 3 & 4 \\ 5 & 6 & 7 & 8 \\ 9 & 10 & 11 & 12 \end{bmatrix}$$

then its transpose would be:

$$\boldsymbol{M}^T = \begin{bmatrix} 1 & 5 & 9 \\ 2 & 6 & 10 \\ 3 & 7 & 11 \\ 4 & 8 & 12 \end{bmatrix}$$

A transposed column vector becomes a row vector, and vice versa:

$$\boldsymbol{v} = \begin{bmatrix} a_1 \\ \vdots \\ a_m \end{bmatrix}$$

$$\boldsymbol{v}^T = \begin{bmatrix} a_1 & \dots & a_m \end{bmatrix}$$

1.1.1.8 A9. Matrix Multiplication

Not all pairs of matrices can be multiplied. Furthermore, matrix multiplication is not commutative, as was matrix addition. In order for matrix ***M*** and ***N*** to be "multipliable" (usually called conformable with respect to multiplication), to obtain the product ***MN***, the number of columns of ***M*** must equal the number of rows in matrix ***N***. Symbolize the entries of matrix ***M*** as a_{ij}, and the entries of matrix ***N*** as b_{jk}. If ***M*** has m rows and n columns, and ***N*** has n rows and r columns, then the product matrix:

$$\boldsymbol{P} = \boldsymbol{MN}$$

will have m rows and r columns. The ith row and jth column of ***P*** will be the sum:

$p_{ij} = a_{i1}{*}b_{1j} + a_{i2}{*}b_{2j} + \ldots + a_{in}{*}b_{nj}$,

As an example, consider:

$$\boldsymbol{M} = \begin{bmatrix} 1 & 2 \\ 3 & 4 \end{bmatrix}$$

$$\boldsymbol{N} = \begin{bmatrix} 5 & 6 & 7 \\ 8 & 9 & 10 \end{bmatrix}$$

Then:

$$\boldsymbol{P} = \boldsymbol{MN} = \begin{bmatrix} 1*5+2*8 & 1*6+2*9 & 1*7+2*10 \\ 3*5+4*8 & 3*6+4*9 & 3*7+4*10 \end{bmatrix} = \begin{bmatrix} 21 & 24 & 27 \\ 47 & 54 & 61 \end{bmatrix}$$

The product ***NM*** is not defined.

A very important special case of matrix multiplication is between an $m \times n$ matrix and an $n \times 1$ vector (which is a very special kind of matrix). That is:

$$\boldsymbol{Ax} = \begin{bmatrix} a_{11} & a_{12} & \cdots & a_{1n} \\ \vdots & \cdots & \ddots & \vdots \\ a_{m1} & a_{m2} & \cdots & a_{mn} \end{bmatrix} \begin{bmatrix} x_1 \\ x_2 \\ \vdots \\ x_n \\ \\ \end{bmatrix} = \begin{bmatrix} b_1 \\ \vdots \\ b_m \end{bmatrix} = \boldsymbol{b}$$

1.1.1.9 A10. Dot or Scalar Product

Two vectors can be multiplied in a fashion that yields a scalar. In order to obtain this scalar product, one vector must be a row vector and the other a column vector, and they both must have equal numbers of components. The "dot" product is symbolized with a "•". If:

$$\boldsymbol{v}^T = \begin{bmatrix} a_1 & \dots & a_n \end{bmatrix}$$

$$\boldsymbol{w} = \begin{bmatrix} b_1 \\ \vdots \\ b_n \end{bmatrix}$$

then the dot product $\boldsymbol{v}^T \cdot \boldsymbol{w}$ is:

$$\boldsymbol{v}^T.\boldsymbol{w} = \begin{bmatrix} a_1 & \dots & a_n \end{bmatrix} \begin{bmatrix} b_1 \\ \vdots \\ b_n \end{bmatrix} = a_1 * b_1 + \dots + a_n * b_{n.}$$

On the other hand, the product $\boldsymbol{w}\boldsymbol{v}^T$ is actually a matrix. Suppose:

$$\boldsymbol{w} = \begin{bmatrix} b_1 \\ \vdots \\ b_m \end{bmatrix}$$

with $m \neq n$.

The vector $\boldsymbol{w}$ can be thought of as a matrix with m rows and 1 column, and $\boldsymbol{v}^T$ can be thought of as a matrix with 1 row and n columns. Thus, the product $\boldsymbol{w}\boldsymbol{v}^T$ is a matrix with m rows and n columns.

A vector has "length," which we will designate as $\|\boldsymbol{v}\|$, and is defined to be:

$$\|\upsilon\| = \sqrt{\sum_{i=1}^{n} a_i^2} = (v^T \cdot v)^{1/2}$$

Inasmuch as this length is a scalar, another vector that has length = 1 can be constructed as:

$$\tilde{\boldsymbol{v}} = \frac{1}{\|\upsilon\|}\upsilon$$

A vector of all 0 s has length = 0.

1.1.1.10 A11. Square Matrices, the Identity Matrix, and Matrix Inverses

A square matrix is one that has the same number of rows and columns. They turn out to be a very important case for a wide variety of applications. For one thing the product of a matrix and its transpose is always a square matrix. If $\boldsymbol{M}$ is an $m \times n$ matrix, then its transpose, $\boldsymbol{M}^T$, is an $n \times m$ matrix, and:

$$\boldsymbol{S}_1 = \boldsymbol{M}^T\boldsymbol{M}.$$

is a square matrix with dimensions $n \times n$, and

$$\boldsymbol{S}_2 = \boldsymbol{M}\boldsymbol{M}^T.$$

is a square matrix with dimensions $m \times m$.

So, going back to our least squares regression, the matrix $\boldsymbol{Z}$ is $n \times k$, so that $\boldsymbol{Z}^T$ is $k \times n$, and therefore $\boldsymbol{Z}^T\boldsymbol{Z}$ is a $k \times k$ square matrix.

An Identity matrix is a special square matrix where most of the entries are 0, except for the "diagonal entries (a_{ii}) which are all 1. In general, the $n \times n$ matrix, $\boldsymbol{I}_n$ is:

$$\boldsymbol{I}_n = \begin{bmatrix} 1 & 0 & \dots & 0 \\ 0 & 1 & \dots & 0 \\ \vdots & 0 & \ddots & \vdots \\ 0 & 0 & \dots & 1 \end{bmatrix}$$

It turns out that for any $n \times n$ square matrix, $\boldsymbol{S}_n$, the products with the identity matrix of corresponding dimensions is the matrix itself:

$$\boldsymbol{S}_n\boldsymbol{I} = \boldsymbol{I}\boldsymbol{S}_n = \boldsymbol{S}_n.$$

This begs the question, can we find another matrix so that its product with square matrix $\boldsymbol{S}_n$ is the $n \times n$ identity? The answer is, sometimes. This matrix, if it exists, is called the inverse of $\boldsymbol{S}_n$. When a square matrix has an inverse, which we denote with a superscript "–1," it has the property that:

$$\boldsymbol{S}_n\boldsymbol{S}_n^{-1} = \boldsymbol{S}_n^{-1}\boldsymbol{S}_n = \boldsymbol{I}_n$$

In the case of multiple regression as described earlier, the answer to the existence of the inverse matrix $[\boldsymbol{Z}^T\boldsymbol{Z}]^{-1}$ is a resounding: almost always. Oh. The usual case for regression is that the matrix:

$$\boldsymbol{S}_k = \boldsymbol{Z}^T\boldsymbol{Z}.$$

Does in fact have an inverse matrix:

$$\boldsymbol{S}_k^{-1} = \left[\boldsymbol{Z}^T\boldsymbol{Z}\right]^{-1}$$

Is it possible to know whether this matrix exists? Generally, software will tell you if you try to fit a multiple regression and the matrix $\boldsymbol{Z}^T\boldsymbol{Z}$ is not invertible. This condition will occur if any column of the regressor matrix, $\boldsymbol{Z}$, is in fact a linear function of other columns. Such a condition is called collinearity. Sometimes a column is very close to being a linear function of other columns, but not quite. Such a condition is called ill-conditioning. Ill conditioned matrices may have an inverse, but it may be too hard or at least very difficult to compute it (we have not discussed how to actually find an inverse at this point; suffice it to say that there is a numerical procedure for doing so). As an example, consider the data in Table A.6.1. In this case, X1 is just the numbers from 1 to 9, and $Y = 2*\text{X1} + 5$ exactly. In the R-code shown in Fig. A.6.1, the variable X2 is X1 plus a little noise (rnorm generates random normal variable values).

The output is shown in Fig. A.6.2

The first regression (linreg1) was perfectly fine. The intercept was close to 5 and the slope parameter (coefficient for X2) was close to 2. Remember that we added some random noise, so that the fit would not be "perfect."

The second regression had the collinear term X3 included. The *lm* function could not estimate its associated coefficient. The message "(1 not defined because of singularities)" indicates that there is some collinearity with respect to variable X3, hence the term "singularities."

As to how a square matrix' inverse is computed, it is accomplished through some variant of a numerical procedure called Gauss-Jordan elimination. The details of this process may be found in any book on linear algebra (see, for example, Strang, 2016).

1.1.1.11 A12. Determinants

Square matrices have a number associated with them called a determinant. Determinants in general are computed using the Gauss-Jordan elimination process and are in fact related to a square matrix' inverse. Perhaps the most important property of the determinant is that if it is 0, then the associated square matrix does not have an inverse. A determinant can either be negative, 0, or positive. The general definition of the process by which a determinant is computed is beyond the scope of this discussion. It is important to realize that such a quantity exists. The determinant

Table A.6.1 Perfect linear relationship example

X1	Y
1	7
2	9
3	11
4	13
5	15
6	17
7	19
8	21
9	23

Fig. A.6.1 R Code for linear regressions with Noisy X and Collinear X

```
setwd("<your path here>")
df1 <- read.csv("20230216 Appendix A.6.1.csv")
#
# Variables:
#   X1
#   Y
#   Note:  Y = 2X + 5, without noise
#
#
attach(df1)
size <- length(X1)
X2 <- X1 + rnorm(n=size,mean=0,sd=.75)
X3 <- 2.5*X2
lnreg0 <- lm(Y ~ X1)
lnreg1 <- lm(Y ~ X2)
lnreg2 <- lm(Y ~ X2 + X3)

summary(lnreg1)
summary(lnreg2)
```

is often symbolized with either the functional notation det(***S***) or “absolute value” bars surrounding the name of the square matrix, e.g.:

$\det(\boldsymbol{S}) = |\boldsymbol{S}|$.

However, a determinant can in fact be negative.

R has a built-in determinant function, det(.), where the argument is a square matrix. Consider this example:

```
> summary(lnreg1)
Call:
lm(formula = Y ~ X2)
Residuals:
   Min      1Q  Median      3Q     Max
-2.2965 -1.6409 -0.3524  1.0603  3.1833
Coefficients:
             Estimate  Std. Error  t value  Pr(>|t|)
(Intercept)     5.407       1.513    3.574  0.009046 **
X2              1.974       0.278    7.102  0.000193 ***
---
Signif. codes:  0 '***' 0.001 '**' 0.01 '*' 0.05 '.' 0.1 ' ' 1
Residual standard error: 2.044 on 7 degrees of freedom
Multiple R-squared:  0.8781,    Adjusted R-squared:  0.8607
F-statistic: 50.44 on 1 and 7 DF,  p-value: 0.0001933
> summary(lnreg2)
Call:
lm(formula = Y ~ X2 + X3)
Residuals:
   Min      1Q  Median      3Q     Max
-2.2965 -1.6409 -0.3524  1.0603  3.1833

Coefficients: (1 not defined because of singularities)
             Estimate  Std. Error  t value  Pr(>|t|)
(Intercept)     5.407       1.513    3.574  0.009046 **
X2              1.974       0.278    7.102  0.000193 ***
X3                 NA          NA       NA        NA
---
Signif. codes:  0 '***' 0.001 '**' 0.01 '*' 0.05 '.' 0.1 ' ' 1
Residual standard error: 2.044 on 7 degrees of freedom
Multiple R-squared:  0.8781,    Adjusted R-squared:  0.8607
F-statistic: 50.44 on 1 and 7 DF,  p-value: 0.0001933
```

Fig. A.6.2 Output of linear regressions

```
> col1 <- c(1,2,3)
> col2 <- c(4,5,6)
> col3 <- c(9,7,8)
> mat1 <- cbind(col1,col2,col3)
> det(mat1)
[1] -9
```

Although no simple formula exists for determinants in general, a 2 × 2 square matrix has a fairly simple determinant formula. For 2 × 2 matrix:

$$\boldsymbol{A} = \begin{bmatrix} a & b \\ c & d \end{bmatrix}$$

the determinant is:

$$\det(\boldsymbol{A}) = ad - bc$$

Another special case of determinant is when the square matrix is symmetric across its diagonal. That is, if a_{ij} represents the element of the matrix in the ith row and jth column, then the matrix is symmetric, also referred to as triangular, if $a_{ij} = a_{ji}$, $i \neq j$. Then the determinant is equal to the product of the diagonal elements:

$$\det(\boldsymbol{A}) = \prod_{i=1}^{n} a_{ii}$$

The matrix A is $n \times n$. This fact is most useful for a special matrix, called the variance-covariance matrix, or sometimes called the covariance matrix. If X_1 and X_2 are random variables with means μ_1 and μ_2, respectively, then the covariance of X_1 and X_2 is:

$$\operatorname{cov}(X_1, X_2) = \operatorname{cov}(X_2, X_1) = \sigma_{12} = \sigma_{21} = \int_{-\infty}^{+\infty}\int_{-\infty}^{+\infty} (x_1 - \mu_1)(x_2 - \mu_2) f(x_1, x_2)\, dx_1 dx_2$$

The covariance between two variables X_1 and X_2 could be estimated from n observations with the formula:

$$C(X_1, X_2) = \frac{1}{n}\sum_{i=1}^{n} (x_{1i} - \overline{x}_1)(x_{2i} - \overline{x}_2)$$

where $\overline{x}$ is the arithmetic average.

The diagonal of the covariance matrix is just the variances of the respective variables. So, for example, the covariance matrix of X_1 and X_2 could be symbolized as:

$$\Sigma = \begin{bmatrix} \sigma_1^2 & \sigma_{12} \\ \sigma_{21} & \sigma_2^2 \end{bmatrix}$$

It is very likely that there will never be a need to compute a determinant per se. Its most obvious application in the area of multivariate analysis, i.e., situations where there are multiple

"response" variables that are related to each other in some fashion. Many of the multivariate methods involve something called eigenvalues and eigenvectors, which are computed using equations with determinants.

1.1.1.12 A13. Eigenvalues and Eigenvectors

Suppose $\boldsymbol{A}$ is a square matrix of dimensions $n \times n$:

$$\boldsymbol{A} = \begin{bmatrix} a_{11} & \cdots & a_{1n} \\ \vdots & \ddots & \vdots \\ a_{n1} & \cdots & a_{nn} \end{bmatrix}$$

Sometimes, especially when the components of matrix $\boldsymbol{A}$ are pairwise correlations between a set of response variables, it is of interest to find a scalar, λ, such that:

$$\boldsymbol{Ax} = \lambda \boldsymbol{x}$$

Since multiplying the identity matrix by a vector yields the original vector:

$$\boldsymbol{Ix} = \boldsymbol{x}$$

then the previous equation can be expressed as:

$$\boldsymbol{Ax} = \lambda \boldsymbol{Ix}$$

Because the two sides of this equation both have the same dimensions ($n \times 1$), the right-hand-side can be subtracted from the left:

$$\boldsymbol{Ax} - \lambda \boldsymbol{Ix} = \boldsymbol{0}$$

where $\boldsymbol{0}$ is an $n \times 1$ vector of all zeros. The vector x can be "factored" to yield:

$$\left(\boldsymbol{A} - \lambda \boldsymbol{I}\right)\boldsymbol{x} = \boldsymbol{0}$$

It turns out that the equation only has a solution if:

$$\left|\boldsymbol{A} - \lambda \boldsymbol{I}\right| = \left|\boldsymbol{0}\right| = 0$$

The determinant of the $n \times 1$ vector of all zeros is 0.

It turns out that there are potentially n different values of λ that satisfy the determinant equation, called the characteristic equation for matrix $\boldsymbol{A}$. That is, the determinant

$$|\boldsymbol{A} - \lambda \boldsymbol{I}|$$

actually is a polynomial of order n, so setting it equal to 0 allows for the possibility of n solutions, or roots. The roots of the characteristic equation are called, believe it or not, characteristic roots, or characteristic values, or eigenvalues. Each root, λ_i then corresponds to a vector, $\boldsymbol{x}_i$, that satisfies the equation:

$$\boldsymbol{A}\boldsymbol{x}_i = \lambda_i \boldsymbol{x}_i$$

The solution vectors, $\boldsymbol{x}_i$, $i = 1, n$, are called characteristic vectors, or eigenvectors.

1.1.1.13 A14. Diagonalization and Powers of Matrices

A diagonal matrix is a square ($n \times n$) matrix with diagonal entries, a_{ii} not necessarily 0, but all other entries necessarily 0 ($a_{ij} = 0$, $i \neq j$). Of particular interest is a diagonal matrix, $\boldsymbol{\Lambda}$, whose diagonal entries are the eigenvalues of another $n \times n$ matrix, $\boldsymbol{A}$. If matrix $\boldsymbol{S}$ is the matrix of eigenvectors of A, then it turns out that (Strang, 2016):

$$\boldsymbol{A} = \boldsymbol{S}\,\Lambda \mathbf{S}^{-1}$$

While there are many applications and mathematical reasons why this relation is important, we will focus here on one fact, relating to powers of a matrix. The kth power of matrix $\boldsymbol{A}$ is simply $\boldsymbol{A}$ multiplied by itself k times. It turns out that for a diagonal matrix, say our matrix of eigenvalues, $\boldsymbol{\Lambda}$, the kth power is simply a diagonal matrix whose diagonal elements are a_{ii}^k. This is very convenient, inasmuch as it makes computing the kth power of $\boldsymbol{A}$ fairly easy:

$$\boldsymbol{A}^k = \boldsymbol{S}\boldsymbol{\Lambda}^k \boldsymbol{S}^{-1} = \boldsymbol{S} \begin{bmatrix} a_{11}^k & \cdots & 0 \\ \vdots & \ddots & \vdots \\ 0 & \cdots & a_{nn}^k \end{bmatrix} \boldsymbol{S}^{-1}$$

All but the first $\boldsymbol{S}$ and last $\boldsymbol{S}^{-1}$ will "cancel" each other, since $\boldsymbol{S}\boldsymbol{S}^{-1} = \boldsymbol{I}$, the identity matrix.

The R function *eigen*() can be used to extract both eigenvalues and eigenvectors from a square matrix. Figure A.14.1 shows some code and associated output for using *eigen*() with a 4 × 4 matrix. The code also shows how the original matrix can be "recovered" using the eigenvalues and eigenvectors. The function *diag*() is used to generate the diagonal matrix from the vector of eigenvalues.

```
setwd("<your path here>")
df1 <- read.csv("20170220 Example A.9 4x4 Matrix.csv")
#
#
# VARIABLES:
# s1
# s2
# s3
# s4
#
attach(df1)
Amat <- as.matrix(df1)
Smat <- eigen(Amat)
Smat$values
Smat$vectors
Lamda <- diag(Smat$values)
Eig <- as.matrix(Smat$vectors)
EigInv <- solve(Eig)
Arecover <- Eig%*%Lamda%*%EigInv
Lamda
Eig
Amat
Arecover
#####
```

Fig. A.14.1 Extracting Eigenvalues and Eigenvectors and Diagonalization

1.1.1.14 A15. Least Squares and Variants

1.1.1.14.1 Solving Linear Equations

When there are *k* equations in *k* unknowns, the problem of solving for the unknowns can be stated in matrix/vector form:

$$\mathbf{y} = \mathbf{X}\beta$$

OUTPUT

```
> Lamda
     [,1]     [,2]        [,3]          [,4]
[1,]    1  0.0000000 0.0000000  0.000000e+00
[2,]    0 -0.4422144 0.0000000  0.000000e+00
[3,]    0  0.0000000 0.1922144  0.000000e+00
[4,]    0  0.0000000 0.0000000 -1.289329e-17
> Eig
     [,1]       [,2]       [,3]        [,4]
[1,]  0.5  0.7026433 -0.1512474 -0.07142857
[2,]  0.5 -0.1993420  0.8144128  0.50000000
[3,]  0.5 -0.3825925  0.1353958 -0.78571429
[4,]  0.5 -0.5658429 -0.5436212  0.35714286
> Amat
      s1  s2   s3   s4
[1,] 0.1 0.2 0.30 0.40
[2,] 0.4 0.3 0.20 0.10
[3,] 0.5 0.2 0.15 0.15
[4,] 0.6 0.1 0.10 0.20
> Arecover
     [,1] [,2] [,3] [,4]
[1,]  0.1  0.2 0.30 0.40
[2,]  0.4  0.3 0.20 0.10
[3,]  0.5  0.2 0.15 0.15
[4,]  0.6  0.1 0.10 0.20
>
```

Fig. BM1.3 (continued)

where y is a $k \times 1$ column vector of known values. $\mathbf{X}$ is a $k \times k$ matrix of coefficients. Finally, $\boldsymbol{\beta}$ is a $k \times 1$ vector of unknowns, for which we would like to solve. Assuming that $\mathbf{X}$ is an invertible matrix, and its inverse is $\mathbf{X}^{-1}$, the solution is simply:

$$\beta = \mathbf{X}^{-1}\mathbf{y}$$

What if X is not a square matrix? Suppose there are more equations, n, than unknowns, k. Then X cannot be simply inverted. However, the problem can still be solved using something called a generalized, or Moore-Penrose inverse (Strang, 1980):

$$\mathbf{X}^{*} = \left[\mathbf{X}^{T}\mathbf{X}\right]^{-1}\mathbf{X}^{T}$$

The matrix $\mathbf{X}^T$ is the transpose of $\mathbf{X}$. So $\mathbf{X}^T\mathbf{X}$ is a $k \times k$ square matrix. The number of columns in X is $k < n$. A set of columns are said to be linearly dependent if one of them can be computed as a linear function of the others. Statistically speaking, two columns in the $\mathbf{X}$ matrix are linearly dependent if the two variables they represent are correlated with each other. If the set of columns are not linearly dependent, they are called linearly independent. The rank of a matrix is the number of its linearly independent columns, also called its dimension. The number of linearly independent columns in $\mathbf{X}$ is k, and the number of linearly independent rows of $\mathbf{X}^T$ is also k. A square matrix is invertible if the number of columns is its rank. This means that the rank of $\mathbf{X}^T\mathbf{X}$ is also k, so that it is invertible. Thus:

$$\mathbf{X}^{*}\mathbf{X} = \left[\mathbf{X}^{T}\mathbf{X}\right]^{-1}\mathbf{X}^{T}\mathbf{X} = \mathbf{I}_{k}$$

The matrix $\mathbf{I}_k$ is the $k \times k$ identity matrix. So finally:

$$\left[\mathbf{X}^{T}\mathbf{X}\right]^{-1}\mathbf{X}^{T}\mathbf{y} = \left[\mathbf{X}^{T}\mathbf{X}\right]^{-1}\mathbf{X}^{T}\mathbf{X}\beta = \mathbf{I}_{k}\beta = \beta$$

It turns out that this is actually the least squares solution to the regression model:

$$\mathbf{y} = \mathbf{X}\beta + \varepsilon$$

where $\boldsymbol{\varepsilon}$ is a $k \times 1$ vector of random "errors."

So, as long as the number of observations, i.e., the length of vector $\mathbf{y}$, n, is greater than or equal to k, the number of regression parameters, the least squares solution exists. What happens if $n < k$? Consider a simplistic example, with $n = 2$ and $k = 3$. Then the matrix equation is still:

$$\mathbf{y} = \mathbf{X}\beta$$

(for now, forget the error vector, $\boldsymbol{\varepsilon}$).

Express the equation in a more expanded form:

$$\begin{bmatrix} y_1 \\ y_2 \end{bmatrix} = \begin{bmatrix} x_{11} & x_{12} & x_{13} \\ x_{21} & x_{22} & x_{23} \end{bmatrix} \begin{bmatrix} \beta_1 \\ \beta_2 \\ \beta_3 \end{bmatrix}$$

Note that the product $\mathbf{X}^T\mathbf{X}$ would be a 3×3 matrix, and $\mathbf{X}^T\mathbf{y}$ would be 3×1 vector:

$$\mathbf{X}^T\mathbf{y} = \begin{bmatrix} x_{11} & x_{21} \\ x_{12} & x_{22} \\ x_{13} & x_{23} \end{bmatrix} \begin{bmatrix} y_1 \\ y_2 \end{bmatrix}$$

So it seems that we could multiply both sides of the original equation by $\mathbf{X}^T$:

$$\mathbf{X}^T\mathbf{y} = \mathbf{X}^T\mathbf{X}\beta$$

The problem is that when $k > n$, the matrix $\mathbf{X}^T\mathbf{X}$ cannot be inverted, even though it is a $k \times k$ square matrix. This is because the $\mathbf{X}$ matrix, having more columns than rows, has some columns that can be expressed as linear functions of the other columns, i.e., they are linearly dependent. The rank of a transpose matrix is equal to the number of linearly independent rows, which is the number of linearly independent columns of the original matrix. Thus, if a matrix, $\mathbf{X}$, is $n \times k$ with $k > n$, then its rank is $k - n$ (assuming there are $k - n$ linearly independent columns in $\mathbf{X}$) and its transpose, $\mathbf{X}^T$, also had rank $k - n$. The product $\mathbf{X}^T\mathbf{X}$ is a $k \times k$ matrix. It turns out that the product of two matrices is a matrix whose rank is the minimum of the two factor matrices, so rank of $\mathbf{X}^T\mathbf{X}$ is $k - n$. However, $\mathbf{X}^T\mathbf{X}$ has k rows and k columns. Since its rank ($k - n$) is less than the number of columns (and rows), it is not invertible. Therefore, there is no least squares solution to a problem with k regressors and n observations where $k > n$.

1.1.1.15 A16. Taylor Series Expansions

Suppose you have a function, $f(x)$, that is nice and smooth, so that it has derivatives. The value of f can be approximated if you know the values of its derivatives (somehow you knew the values of its derivatives but not the function itself).

So, it turns out that the Taylor series approximation for $f(x)$ is:

$$f(x) \approx f(0) + f'(0)x + \frac{1}{2!}f''(0)x^2 + \frac{1}{3!}f'''(0)x^3 + \frac{1}{4!}f''''(0)x^4 \ldots$$

So, if you know the function's value when $x = 0$, call it $f(0)$, and the value of some arbitrary number of its derivatives at $x = 0$, you could approximate the value of the function at any arbitrary value of x. The more terms you add, the better the approximation.

As an example, suppose:

$$f(x) = 5\sin(2x)$$

That means:

$$f'(x) = 2 * 5\cos(2x)$$

$$f''(x) = -2 * 2 * 5\sin(2x)$$

$$f'''(x) = -2 * 2 * 2 * 5\cos(2x)$$

$$f''''(x) = 2 * 2 * 2 * 2 * 5\sin(2x)$$

We also can compute:

$$f(0) = 5\sin(0) = 0$$

$$f'(0) = 2 * 5\cos(0) = 10$$

$$f''(0) = -2 * 2 * 5\sin(0) = 0$$

$$f'''(0) = -2 * 2 * 2 * 5\cos(0) = -40$$

$$f''''(x) = 2 * 2 * 2 * 2 * 5\sin(0) = 0$$

So an approximation of the function:

$$f(x) = 5\sin(2x)$$

is:

$$f(x) \approx f(0) + f'(0)x + \frac{1}{2!}f''(0)x^2 + \frac{1}{3!}f'''(0)x^3 + \frac{1}{4!}f''''(0)x^4 =$$

$$f(x) \approx 0 + 10x + 0 + \frac{1}{3!}(-40)x^3 + 0$$

So we know that $\sin\left(2 * \frac{\pi}{2}\right) = 0$, and $f(x) = 5\sin\left(2\frac{\pi}{2}\right) = 0$. Our approximation is:

$$f\left(\frac{\pi}{2}\right) \approx 0 + 10\left(\frac{\pi}{2}\right) + 0 + \frac{1}{3!}(-40)\left(\frac{\pi}{2}\right)^3 + 0 \approx -10.1306$$

OK, so it's not a very good approximation. I guess we would need to use more terms to get a better approximation.

1.1.1.16 A17. Quadratic Solution and Vertices

The solution to the equation:

$$ax^2 + bx + c = 0$$

is:

$$x = \frac{-b \pm \sqrt{b^2 - 4ac}}{2a}$$

That is, the value of x that makes the quadratic equal to zero is either:

$$x_1 = \frac{-b - \sqrt{b^2 - 4ac}}{2a}$$

or

$$x_2 = \frac{-b + \sqrt{b^2 - 4ac}}{2a}$$

Sometimes these values are complex numbers.
As an example, suppose we have the quadratic expression:

$$2x^2 + 1x - 6$$

The solution to the equation:

$$2x^2 + 1x - 6 = 0$$

Is:

$$x = \frac{-1 \pm \sqrt{1^2 - 4(2)(-6)}}{2(2)} = \frac{-1 \pm \sqrt{49}}{4} = \left(\frac{-8}{4}, \frac{6}{4}\right)$$

If you put either −8/4 or 6/4 into the expression, you will get 0.

1.1.1.17 A18. Vertex

The vertex of a parabola (which is described by a quadratic expression) can be derived using derivatives.

Let:

$$f(x) = ax^2 + bx + c$$

From the rules of differentiation, we get:

$$f'(x) = 2ax + b$$

To find the value of x at which the original function is a maximum (for down-pointing parabolas) or minimum (for up-pointing parabolas) set the derivative (also called the first derivative) to zero and solve for x:

$$f'(x_v) = 2ax_v + b = 0$$

$$x_v = \frac{-b}{2a}$$

To find the value of the parabola at the vertex, simply put the value of x_v back in the original expression.

For example, suppose:

$$f(x) = x^2 - 2x + 7$$

Then:

$$f'(x) = 2x - 2$$

Setting this equal to 0 and solving:

$$2x_v - 2 = 0$$

$$x_v = \frac{2}{2} = 1$$

The vertex is at $x_v = 1$. The parabola's height at $x_v = 1$ is:

$$f(1) = 1^2 - 2(1) + 7 = 6$$

The way to know if the parabola is pointing up (so that the vertex is its minimum) or pointing down (so that the vertex is its maximum) is simple. If $a > 0$, the parabola is pointing up. If $a < 0$, the parabola is pointing down. So, in this example, since $a = 1 > 0$, the value 6 is the parabola's minimum.

1.1.1.18 A19. Inequalities and Absolute Value Expressions

An inequality expression is just like an equation, with the "=" sign replaced with one of the following:

< (strictly less than)
>(strictly greater than)
≤ (less than or equal to)
≥ (greater than or equal to)
So the expressions:
A < B means "A is strictly less than B".
A > B means "A is strictly greater than B".
A ≤ B means "A is either less than or equal to B".
A ≥ B means "A is either greater than or equal to B".

OK, so those are the elementary definitions. It gets a little more complicated when A or B are algebraic expressions in terms of some unknown variables, say x or y. Sometimes we are asked to "solve" an inequality for the variable. The solving process is very similar to solving equalities, with one very important exception. If you multiply or divide both sides of the inequality by a negative number, the direction of the inequality changes. This is true whether the inequality sign is strictly less than, less than or equal, strictly greater than, or greater than or equal. Furthermore, the "solution" is often an infinite set of numbers, instead of one (in linear equalities) or two (quadratic equalities) numbers.

For example, consider the inequality:

$3x - 6 < 0$

If this were the equality:

$3x - 6 = 0.$

We would solve for x by adding 6 to both sides and dividing both sides by 3:

$3x = 6$
$x = 6/3 = 2.$

Do the same thing for the inequality:

$3x < 6$
$x < 2.$

So, any value of x that is strictly less than 2 will satisfy the inequality:

$3x - 6 < 0$

The biggest difference in solving inequalities is when multiplying or dividing both sides by a negative number. In that case, the direction of the inequality is reversed. For example:

$-3x \geq 6$

Dividing both sides by -3 gives you:

$x \leq 6/(-3)$,

or

$x \leq -2$,

Notice that it does not matter if the inequality is strict ($<$ or $>$) or not strict ($\leq$ or $\geq$).

So far the inequalities only involved linear terms. What about quadratic expressions?

Suppose you had:

$3x^2 - 27 > 0$

We could use the quadratic formula, but this expression is too simple. Add 27 to both sides and divide by 3:

$x^2 > 9$,

You can take square roots of both sides (this will NOT change the direction of the inequality), but keep in mind that the solution can be both negative or positive:

$x > \pm\sqrt{9} = \pm 3$,

OK, this does not make sense. Suppose, for example, $x = -2 > -3$. Well:

$x^2 = (-2)^2 = 4$ which is certainly not >9. How do we resolve this? The answer is absolute value.

If $x^2 > 9$, then clearly $x > 3$ makes the statement true, but so does $x < -3$. In other words, the solution is:

$|x| > 3$.

So, whenever you have a squared variable, think of it as:

$|x|^2$

As another example, suppose:

$5x^2 > 125$

First translate this expression into something with an absolute value:

$5|x|^2 > 125$

Then divide both sides by 5:

$|x|^2 > 25$

Now take square roots of both sides:

$|x| > 5$

So this means any value of x that is either greater than +5 or less than −5 will make the inequality true.

Notice that the direction of the inequality did not change, since we never multiplied or divided both sides by a negative number.

If the original inequality were < instead of >:

$5x^2 < 125$

The solution would be:

$|x| < 5$

This means that if x is anywhere between −5 and +5, the inequality is true. Since this is a strict inequality, the values $x = -5$ or $x = +5$ do not make the inequality true (i.e., −5 and +5 do not satisfy the inequality).

1.1.1.19 A20. Expressions with Absolute Values

Sometimes you may encounter an expression that is linear in terms of the variable, but also has an absolute value for the expression. For example:

$|2x - 8| < 10$

Such inequalities actually are describing two separate expressions:

$2x - 8 < 10$

and

$-(2x - 8) < 10$

This is because

$|-(2x - 8)| = |2x - 8|$

So, get a complete answer to the question:

What values of x will make the inequality true?

You must solve both inequalities.

The first one is fairly straight-forward:

$2x < 10 + 8 = 18$
$x < 18/2$

so, $x < 9$.

The other inequality requires multiplication by a negative number (−1):

$-(2x - 8) < 10$
$2x - 8 > -10$ (notice the change of direction in the inequality sign)

$2x > -10 + 8 = -2$
$x > -1$ (notice that the direction of the inequality did NOT change when dividing by 2)

So, the complete answer is that $x > -1$ and $x < 9$. To see if this is correct, substitute for x any number between -1 and 9 (but don't include either -1 or 9, since this is a strict inequality.)

Try $x = 0$:

$|2(0) - 8| = 8 < 10$

Or try $x = 5$:

$|2(5) - 8| = 2 < 10.$

One more; try $x = -1/2$:

$|2(-1/2) - 8| = |-1 - 8| = |-9| = 9 < 10$

The case of absolute values with quadratic expressions is very similar in nature, namely that the absolute value of the expression generates two inequalities to solve. For example:

$|x^2 - 2| \leq 1$

The two inequalities are:

$x^2 - 2 \leq 1$

and

$-(x^2 - 2) \leq 1$

In the case of the first, the solution is fairly simple to obtain:

$x^2 - 2 \leq 1$
$x^2 \leq 1 + 2 = 3$

Now that we have:

$x^2 \leq 3$

we must recall how we solved inequalities with squared variables:

$|x|^2 \leq 3$
$|x| \leq \sqrt{3}$

So either

$x \leq \sqrt{3}$

Or

$-x \leq \sqrt{3}$

If $-x \leq \sqrt{3}$, then $x \geq -\sqrt{3}$

So for the first part of the original inequality, either $x \geq -\sqrt{3}$ or $x \leq +\sqrt{3}$. The second part proceeds similarly.

$-(x^2 - 2) \leq 1$
$x^2 - 2 \geq -1,$
$x^2 \geq -1 + 2 = 1.$

so

$|x|^2 \geq 1$

This means that either

$x \geq 1,$

Or

$-x \geq 1$ (i.e., $x \leq -1$)

So in summary, we now have that in order for the original inequality to be true, then:

$x \geq -\sqrt{3}$ or $x \leq +\sqrt{3},$

and

$x \geq 1$ or $x \leq -1,$

Well, if $x = -\sqrt{4}$, it is certainly ≤ -1, but it is NOT $\geq -\sqrt{3}$. Similarly, if $x = +\sqrt{4}$, then it is ≥ 1 but NOT $\leq +\sqrt{3}$.

As a check, let $x = -\sqrt{4}$ in the original inequality:

$|(-\sqrt{4})^2 - 2| = |4 - 2| = 2 > 1$, so $-\sqrt{4}$ is not a solution.

Conversely, if $x = -\sqrt{3}$, then

$|(-\sqrt{3})^2 - 2| = |3-2| = 1$

Since the original inequality is not strict, $-\sqrt{3}$ is in fact a valid solution. In this case, the second part of solving the inequality does not result in any more solutions.

In summary:

1. Inequalities are for the most part solved just like equalities, with one exception;
2. The exception is that when multiplying or dividing both sides by a negative number, the direction of the inequality changes;
3. When the expression has a squared variable, then before solving, convert the squared variable into the absolute value of the squared variable; x^2 becomes $|x|^2$;
4. When an absolute value of the variable is in the expression, then it creates two separate expressions that both must be solved; That is because $|x| = |-x|$;
5. Expressions with absolute values of quadratic expressions may have up to 4 solutions;

1.1.1.20 A21. Complex Numbers and Variables

The imaginary number is symbolized by

$$j = \sqrt{-1}$$

A general imaginary number is a real number, y multiplied by j:

jy

A complex number is a sort of "sum" of a real number and an imaginary number:

$$z = x + jy$$

The "+" is not quite the same as the "+" used to add real numbers; that is, $x + jy$ does not mean the same thing as $x + y$.

There are rules for combining complex numbers in such ways that will be called "addition" and "multiplication." When adding or multiplying complex numbers, you always get another complex number. Real numbers are special cases of complex numbers:

$$z = x + j0 = x$$

Typically, x is referred to as the **Re**al part, and y is the **Im**aginary part, of the complex number:

$$z = x + jy$$

So, if $z_1 = x_1 + jy_1$ and $z_2 = x_2 + jy_2$, then:

$z_1 * z_2 = x_1x_2 + jx_1y_2 + jx_2y_1 - y_1y_2 = (x_1x_2 - y_1y_2) + j(x_1y_2 + x_2y_1)$,
$z_1 \oplus z_2 = x_1 + x_2 + j(y_1 + y_2)$.

The symbols used for multiplication (*) and addition ($\oplus$) are different to indicate that these are for complex numbers. Note that adding or multiplying two complex numbers results in another complex number. Recall that real numbers can be thought of a complex n umbers with $y = 0$, and imaginary numbers can be thought of as complex numbers with $x = 0$. So in that sense complex arithmetic is a generalization of real arithmetic.

A point in Cartesian coordinates (x, y) can be represented in polar coordinates as:

$$x = r\cos\theta$$

$$y = r\sin\theta$$

The polar expression of the point is (r, θ). The value of r, called the modulus, is:

$$r = \sqrt{x^2 + y^2}$$

The value of θ is the angle that a ray from the origin (0, 0) to the point (x, y) makes with the x-axis. It can be computed if you know (x, y) as:

$$\theta = \tan^{-1} \frac{y}{x}$$

Of course, as long as $x \neq 0$. When $x = 0$, $\theta = \frac{\pi}{2} = 90^\circ$.

A complex number, $z = x + jy$, can be expressed in terms of (r, θ) by substituting the polar representations for x and y:

$$z = x + jy = r\left(\cos\theta + j\sin\theta\right) = re^{j\theta}$$

The exponential form is based on Euler's relation:

$$e^{j\theta} = \cos\theta + j\sin\theta$$

It turns out that the "polar" form of a complex number is very useful when performing operations and computing functions of complex numbers.

An important concept is the complex conjugate. If:

$$z = x + jy$$

then its complex conjugate is:

$$\overline{z} = x - jy = re^{-jy}$$

It seems simple enough. Here is an interest set of facts:

$$z\overline{z} = \overline{z}z = \left(x + jy\right)\left(x - jy\right) = x^2 + y^2 = r^2$$

$$\frac{z}{\overline{z}} = \frac{re^{j\theta}}{re^{-j\theta}} = e^{2j\theta}$$

$$\frac{\overline{z}}{z} = \frac{re^{-j\theta}}{re^{j\theta}} = e^{-2j\theta}$$

$$z^k = \left(re^{j\theta}\right)^k = r^k e^{jk\theta}$$

1.1.1.21 A22. Fourier Series

Any piece-wise continuously-valued, periodic function, with period T, can be represented by a series:

$$f(t) = \frac{a_0}{2} + \sum_{i=1}^{\infty} a_i \cos \omega_i t + b_i \sin \omega_i t$$

The values a_0, a_i, b_i are called "Fourier coefficients" and the

$$\omega_i = \frac{2\pi i}{T}$$

are called frequencies. The value:

$$\omega_1 = \frac{2\pi}{T}$$

is sometimes referred to as the principal frequency. Generally, the higher the value of the frequency, the less impact it has on the value of the function $f(t)$.

The series representation of $f(t)$ is called the Fourier Series of $f(t)$. The values ω_i are called "frequencies." Using Euler's relation, the Fourier series can also be expressed as:

$$f(t) = \sum_{k=-\infty}^{+\infty} c_k e^{2\pi j \omega_k t}$$

The coefficients c_k are related to a_k and b_k as:

$$c_k = \frac{a_k - jb_k}{2}$$

Typically, c_{-k} is the complex conjugate of c_k:

$$c_{-k} = \frac{a_k + jb_k}{2}$$

The values c_k are also called "Fourier coefficients," but they have complex values. If you know two values of c_k, say c_1 and c_{-1}, you can solve for a_k and b_k.

It turns out that the Fourier coefficients can be estimated using data and least squares. Such analyses fall under the general topic of time series analysis. In particular, using a time sequence of data, z_1, z_2, …, zt,… z_T, the Fourier coefficients can be estimated up to $\frac{T}{2}$ terms, which, for large enough T, should give a pretty good estimate of the function $z(t)$. In this context, we are using the letter z to represent real numbers (sorry for the confusion).

1.1.1.22 A23. Spectral Analysis of Time Series

Suppose a series of values (e.g., signal amplitudes) are observed at discrete times, $t = 1, 2, \ldots, T$. Represent these values with $z_1, z_2, \ldots, z_T$.

The Fourier model for this series can be expressed as:

$$z_t = \alpha_0 + \sum_{i=1}^{q} \alpha_i \cos(2\pi f_i t) + \beta_i \sin(2\pi f_i t) + \epsilon_t$$

For $t = 1, 2, \ldots, T$

$$q = \begin{cases} \dfrac{T-1}{2}, T \text{ odd} \\ \\ \dfrac{T}{2}, T \text{ even} \end{cases}$$

$$f_i = \frac{i}{T}$$

The values α_0, α_i, and β_i are unknown constants (to be estimated with data via least squares, for example), and ϵ_t is Gaussian white noise (i.e., noise distributed as $N(0, \sigma^2)$) The f_i are referred to as the ith harmonic of the fundamental frequency, $\frac{1}{T}$.

The least squares estimates for the unknown constants (called Fourier coefficients) are given by the formulas:

$$\hat{\alpha}_0 = \bar{z} = \frac{1}{T} \sum_{t=1}^{T} z_t$$

$$\hat{\alpha}_i = \frac{2}{T} \sum_{t=1}^{T} z_t \cos(2\pi f_i t)$$

$$\hat{\beta}_i = \frac{2}{T} \sum_{t=1}^{T} z_t \sin(2\pi f_i t)$$

$i = 1, q$.

If T is even, the estimates for the last coefficients ($i = q$) are a little funny. They are:

$$\hat{\alpha}_q = \sum_{t=1}^{T} (-1)^t z_t$$

and

$$\hat{\beta}_q = 0$$

Note that in the Fourier model, the sum is over the harmonics, f_i, $i = 1, q$, but in the least squares formulas, the sum is over the T observations, z_t, $t = 1, T$. That is, there are $q + 1$ parameters, or coefficients, to estimate, and each one is estimated using all T observations.

Now compute the function, called the sample intensity at f_i, as:

$$\hat{I}(f_i) = \frac{T}{2}\left(\hat{\alpha}_i^2 + \hat{\beta}_i^2\right),\ i = 1, q$$

The plot of $\hat{I}(i)$ on the y-axis and f_i on the x-axis, $i = 1, q$, is called the periodogram, or sample spectrum. If instead of computing the estimates of the coefficients using f_i, you compute them with an arbitrary sequence of values for f between 0 and 0.5, then the plot is called the spectrum instead of the periodogram.

It turns out that the sample intensity and spectrum have a nice relationship to something called the autocovariance (or autocorrelation) function of the series z_t, $t = 1, T$.

Suppose that z_t is "centered", by subtracting the mean value of the time series y_t:

$z_t = y_t - \mathrm{E}(y_t)$,

Then the autocovariance function for z_t is:

$\gamma(h) = \mathrm{E}(z_t z_{t-h})$.

Note that you could use z_{t+h} or just make sure that $t - h \geq 0$, if you like dealing with non-negative time subscripts.

So, the spectrum is the Fourier transform of the autocovariance. More typically, we use the "cosine" Fourier transform:

$$I(f) = \frac{1}{2\pi}\left(\gamma(0) + \sum_{h=1}^{\infty}\gamma(h)\cos(hf)\right)$$

When using sample data, assume that T values of the series have been observed so that:

$$\hat{I}(f_i) = \frac{1}{2\pi}\left(\hat{\gamma}(0) + \sum_{h=1}^{T/2}\hat{\gamma}(h)\cos(hf_i)\right)$$

The values of f_i are taken to be values between 0 and 0.5. The values can either be arbitrarily assigned with uniform increments or computed as described earlier:

$$f_i = \frac{i}{T},\ i = 1, \frac{T}{2}$$

1.1.1.23 A24. A Note on Importing Data

You can either use the reads.csv() function from Base R or the read_excel() function in library readxl. The structure of the data is the same, namely the first row is column headings (variable names), and each row thereafter is a record.

1.2 Appendix B: R Scripts

1.2.1 Chapter 4

1.2.1.23.1 Code 4.1

```
setwd("Your path here")
#
p <- seq(from=0.790,to=0.999,by=0.001)
n.size <- 92
n.1 <- 100
crit.val <- 83
crit.1 <- 90
n.txt <- round(n.size,0)
crit.txt <- round(crit.val,0)
prob.pass.0 <- 1 - pbinom(q=crit.val,size=n.size,prob=p)
prob.fail.0 <- 1- prob.pass.0
AQL.0 <- mean(p[prob.pass.0>=0.94 & prob.pass.0<=0.96])
RQL.0 <- mean(p[prob.pass.0>=0.04 & prob.pass.0<=0.06])
AQL.0.text <- as.character(round(AQL.0,3))
RQL.0.text <- as.character(round(RQL.0,3))

prob.pass.1 <- 1 - pbinom(q=crit.1,size=n.1,prob=p)
AQL.1 <- mean(p[prob.pass.1>=0.94 & prob.pass.1<=0.96])
RQL.1 <- mean(p[prob.pass.1>=0.04 & prob.pass.1<=0.06])
AQL.1.text <- as.character(round(AQL.1,2))
RQL.1.text <- as.character(round(RQL.1,2))
#
plot(x=p,y=prob.pass.0,main="Power Curve",xlab="Pr{Success}"
,type="l",
     ylab="Pr{At least crit.val Successes}",xlim=c(0.73,1.0))
text(x=0.79,y=0.975,labels="Pr{PASS} = 95%")
text(x=0.79,y=0.075,labels="Pr{PASS} = 5%")
text(x=0.79,y=0.80,labels="n = ")
text(x=0.82,y=0.80,labels=n.txt)
text(x=0.777,y=0.70,labels="crit.val = ")
text(x=0.82,y=0.70,labels=crit.txt)
text(x=0.97,y=0.20,labels=AQL.0.text)
text(x=0.82,y=0.20,labels=RQL.0.text)
abline(h=0.95)
abline(h=0.05)
abline(v=RQL.0)
abline(v=AQL.0)
#
```

```
n.2 <- 92
crit.2 <- 83
prob.pass.2 <- 1 - pbinom(q=crit.2,size=n.2,prob=p)
AQL.2 <- mean(p[prob.pass.2>=0.94 & prob.pass.2<=0.96])
RQL.2 <- mean(p[prob.pass.2>=0.04 & prob.pass.2<=0.06])
AQL.2.text <- as.character(round(AQL.2,3))
RQL.2.text <- as.character(round(RQL.2,3))
#
n.3 <- 50
crit.3 <- 45
prob.pass.3 <- 1 - pbinom(q=crit.3,size=n.3,prob=p)
AQL.3 <- mean(p[prob.pass.3>=0.94 & prob.pass.3<=0.96])
RQL.3 <- mean(p[prob.pass.3>=0.04 & prob.pass.3<=0.06])
AQL.3.text <- as.character(round(AQL.3,3))
RQL.3.text <- as.character(round(RQL.3,3))

dev.new()
plot(x=p,y=prob.pass.1,main="Comparing Power Curves",xlab="Pr{Suc
cess}",type="l",
     lty=1,ylab="Pr{At least crit.val
Successes}",xlim=c(0.73,1.0))
points(x=p,y=prob.pass.2,type="l",lty=3)
points(x=p,y=prob.pass.3,type="l",lty=2)
legend(x=0.74,y=0.925,legend=c("n=100","n=92","n=50"),lt
y=c(1,2,3))
abline(h=0.95)
abline(h=0.05)
abline(v=RQL.1,lty=1)
abline(v=AQL.1,lty=1)
abline(v=RQL.2+0.0005,lty=2)
abline(v=AQL.2,lty=2)
abline(v=RQL.3,lty=3)
abline(v=AQL.3,lty=3)

text(x=0.765,y=0.975,labels="Pr{PASS}=95%")
text(x=0.765,y=0.075,labels="Pr{PASS}=5%")
dev.new()
plot(x=p,y=prob.fail.0,main="Power Curve for Inverse Plan",
     xlab="Pr{Success}",ylab="Pr{FAIL}",type="l")
AQL.4 <- RQL.0
RQL.4 <- AQL.0
AQL.4.text <- RQL.0.text
RQL.4.text <- AQL.0.text
abline(h=0.95)
abline(h=0.05)
```

```
abline(v=AQL.4)
abline(v=RQL.4)
text(x=0.90,y=0.975,labels="Pr{FAIL}=95%")
text(x=0.90,y=0.075,labels="Pr{FAIL}=5%")
text(x=0.82,y=0.50,labels=AQL.4.text)
text(x=0.97,y=0.50,labels=RQL.4.text)
```

1.2.1.23.2 Code 4.2

```
setwd("Your path here")
#
N <- 475
n.size <- 100
p.success <- 0.95
D <- round(p.success*N,0)
#
# N = population size
# n.size = sample size
# p.success = p0, null probability of success
# x.crit = critical value for number of successes in the sample
# D.alt = round(p.alt*N,0) = alternative number of successes in N
#
p.alt <- seq(from = 0.75,to = 0.99,by=0.01)
D.alt <- round(p.alt*N,0)
x.crit <- 92
power <- 100*(1-phyper(q=x.crit-1,m=D.alt,n=N-D.alt,k=n.size))
power.B <- 100*(1-pbinom(q=x.crit-1,size=n.size,prob=p.alt))
#
# m = D.alt = number of "white" balls (successes) in N
# n = N - D.alt = number of "black" balls (non-successes) in N
# k = n.size = sample size
#
#  The parameter of interest is D.alt = p.alt*N
#  so p.alt = alternative probability of success in N
#
plot(x=p.alt,y=power,main="Power Curve for Hypergeometric
Sampling Plan",
     type="b",xlab="Pr{Success}",ylab="Power (%)")
abline(h=95)
abline(v=p.success)
text(x=0.80,y=97.5,labels="Pr{PASS}=95%")
dev.new()
plot(x=p.alt,y=power,main="Power Curves - Hypergeometric and
Binomial",
```

```
     type="b",xlab="Pr{Success}",ylab="Power (%)",pch=1,lty=1)
points(x=p.alt,y=power.B,pch=2,lty=2)
abline(h=95)
abline(v=0.95)
text(x=0.80,y=97.5,labels="Pr{PASS}=95%")
legend(x=0.78,y=80,legend=c("Hypergeometric","Binomial"),pch=c(1
,2),lty=c(1,2))
```

1.2.1.1 Chapter 5

1.2.1.1.1 Code 5.1

```
setwd("Your path here")
#
mu <- 100
sigma <- 10
mu.txt <- as.character(round(mu,0))
sd.txt <- as.character(round(sigma,0))
Min.X <- mu - 5.5*sigma
Max.X <- mu + 5.5*sigma
Incr <- (Max.X - Min.X)/70
X.vals <- seq(from=Min.X,to=Max.X,by=Incr)
norm.dens <- 100*dnorm(x=X.vals,mean=mu,sd=sigma)
plot(x=X.vals,y=norm.dens,main="Normal
Density",xlab="X",ylab="Density (%)",
     type="l",lty=1,lwd=3)
abline(v=mu)
text(x=55,y=3,labels="mu = ")
text(x=67,y=3,labels=mu.txt)
text(x=55,y=2.5,labels="sigma = ")
text(x=67,y=2.5,labels=sd.txt)
height.1 <- 100*dnorm(x=mu+sigma,mean=mu,sd=sigma)
segments(x0=mu,y0=height.1,x1=mu+sigma,y1=height.1,lty=1,lwd=3)
#
shape.X <- 26
shape.txt <- as.character(round(shape.X,0))
scale.X <- 3.85
scale.txt <- as.character(round(scale.X,2))
mean.gamma <- shape.X*scale.X
Min.gamma <- qgamma(p=0.001,shape=shape.X,scale=scale.X)
Max.gamma <- qgamma(p=0.999,shape=shape.X,scale=scale.X)
X.gamma <- seq(from=Min.gamma,to=Max.gamma,by=0.001)
gamma.dens <- 100*dgamma(x=X.gamma,shape=shape.X,scale=scale.X)
dev.new()
plot(x=X.gamma,y=gamma.dens,main="Gamma Density",type="l",lwd=3,
```

```
     xlab="X",ylab="Density (%)")
abline(v=mean.gamma)
text(x=130,y=1.8,labels="shape =")
text(x=144,y=1.8,labels=shape.txt)
text(x=130,y=1.6,labels="scale =")
text(x=144,y=1.6,labels=scale.txt)

#
p1 <- pnorm(q=mu+1*sigma,mean=mu,sd=sigma) - pnorm(q=mu-1*sigma,m
ean=mu,sd=sigma)
p2 <- pnorm(q=mu+2*sigma,mean=mu,sd=sigma) - pnorm(q=mu-2*sigma,m
ean=mu,sd=sigma)
p3 <- pnorm(q=mu+3*sigma,mean=mu,sd=sigma) - pnorm(q=mu-3*sigma,m
ean=mu,sd=sigma)
p1
p2
p3
```

1.2.1.2 Chapter 6

1.2.1.2.1 Code 6.1

```
setwd("Your path here")


n <- 30
n.txt <- as.character(round(n,0))
n.tests <- 1
alpha_prime <- 0.05/n.tests
#
# The Cpk hypothesis test:
# H0: Cpk < K_list
# H1: Cpk >= K_list
#
prob_inspec <- seq(from=0.9390,to=0.9999,by=0.0005)
Z0 <- qnorm(prob_inspec)
K_list <- Z0 / 3
Cpk_crit <- 0.8041
Cpk_crit.txt <- as.character(round(Cpk_crit,4))

#
#
#  This is the main part of the code.
#  crit_val_Cpk is the crticial value (minimum sample
#  Cpk required to reject H0 with 95% probabilty
```

```
#  if Cpk = K0)
#
noncenT <- Cpk_crit*3*sqrt(n)
crit_val_Cpk <- noncenT / (3*sqrt(n))
#

prob_reject <- pt(noncenT,df=n-1,ncp=3*sqrt(n)*K_list)


prob_accept <- 1 - prob_reject
n.bin <- 117
n.bin.txt <- as.character(round(n.bin,0))
crit.bin <- 116
prob_accept.bin <- 1-pbinom(crit.bin-1,size=n.
bin,prob=prob_inspec)
Cpk_0
<- round(K_list[mean(which(prob_accept>=0.945 & prob_
accept<=0.955))],4)
Cpk_0.txt <- as.character(round(Cpk_0,4))
Cpk_1
<- round(K_list[mean(which(prob_accept>=0.045 & prob_
accept<=0.055))],4)
Cpk_1.txt <- as.character(round(Cpk_1,4))

plot(x=prob_inspec,y=prob_accept,main="Power Curves",
     xlab="Pr{L<=X<=U}",ylab="Pr{PASS}",pch=1,type="l",lty=1)
points(x=prob_inspec,y=prob_accept.bin,pch=2,type="l",lty=3)
abline(h=0.95)
abline(h=0.05)
legend(x=0.95,y=0.80,legend=c("Cpk Plan","Binomial
Plan"),lty=c(1,3))
text(x=0.947,y=0.975,labels="Pr{PASS}=95%")
text(x=0.947,y=0.075,labels="Pr{PASS} = 5%")
text(x=0.95,y=0.50,labels="n(Cpk) =")
text(x=0.96,y=0.50,labels=n.txt)
text(x=0.95,y=0.40,labels="n(Bin) =")
text(x=0.96,y=0.40,labels=n.bin.txt)

#
prob_within <- 1 - 2*pnorm(-3*K_list)
df_Cpk <- cbind(n,Cpk_crit,prob_inspec,Z0,K_list,prob_
accept,prob_reject)

#compute approximate lower confidence limit for Cpk, given sample
Cpk = K_list
CpkLim <- K_list - qnorm(.95)*sqrt((1/(9*n)) + (K_list**2)/
```

```
(2*(n-1)))
#
# prob_accept = probability of PASSing for Cpk plan
#
write.csv(df_Cpk,"20240630 Cpk OC Curve Naive Crit. Val.
Output.csv")
```

1.2.1.3 Code 6.2

```
setwd("Your path here ")
df1 <- read.csv("20240922 Chapter 6 RMSE Parameters.csv")
#
# this dataset has variables:
# tau = target value for X
# L = lower spec limit
# U = upper spec limit
# pL = Pr{X<L}
# pU = Pr{X>U}
# n = sample size for critical value
#
# It is assumed that the measurement, X, is normally distributed
#
alpha <- 0.05
attach(df1)
n.txt <- as.character(round(n,0))
#C0 <- qnorm(p=1-(pL+pU))/3
#C0 <- 1.0
ZL <- qnorm(pL)# = (L - mustar)/sigstar
ZU <- qnorm(1-pU) # = (U - mustar)/sigstar
# Since pL and pU are given, ZL and ZU are also given.
# They are used to solve for mustar and sigstar.
# mustar and sigstar are the values of E[X] and SQRT(V[X])
# that yield:
# Pr{X<L} = pL and Pr{X>U} = pU exactly
#
# Note that the events {X<L} and {X>U} are mutually exclusive
# so Pr{X<L OR X>U} = pL + pU.
#
#
mustar <- round((L*ZU - U*ZL) / (ZU - ZL),4)
sigstar <- round((U - L) / (ZU - ZL),4)
Ztau <- (tau - mustar)/sigstar #This is the z-score for the
target value, tau.
P.LU <- 1 - (pL + pU)
C0 <- qnorm(p=(1+P.LU)/2)/3
```

```
C0
lamda <- n*(Ztau**2) #non-centrality parameter = 0 when tau
= mustar
# By symmetric, we mean that L and U are symmetrically placed
around tau
# and pL = pU.
#
#nu <- ((n + lamda)**2)/(n+2*lamda)
Cpm.crit <- sqrt(n/qchisq(p=1-alpha,df=n,ncp=lamda))*C0
Ztau.alt <- seq(from=-5.0,to=5.0,by=0.01)
lamda.alt <- n*Ztau.alt**2
#nu.alt <- ((n+lamda.alt)**2)/(n+2*lamda.alt)
n.txt <- as.character(n)

prob.pass <- pchisq(n*(C0/Cpm.crit)**2,df=n,ncp=lamda.alt)
Ztau.05 <- mean(Ztau.alt[prob.pass>=0.04 & prob.pass <= 0.06 &
Ztau.alt > 0.0])
plot(x=Ztau.alt,y=prob.pass,main="Power for Cpm",
     xlab="Ztau",ylab="Pr{PASS}",xlim=c(-6,6))
abline(v=Ztau)
abline(h=0.95)
abline(h=0.05)
text(x=-3.5,y=0.920,labels="Pr{PASS}=95%")
text(x=-3.5,y=0.075,labels="Pr{PASS}=5%")
text(x=-3.5,y=0.85,labels="n = ")
text(x=-3.0,y=0.85,labels=n.txt)
#abline(v=Ztau.05)
#abline(v=-Ztau.05)
df.out <- data.frame(cbind(Ztau.alt,prob.pass))
#df2 <- cbind(tau,L,U,pL,pU,mustar,sigstar,RMSEcrit,ERMSE)
n
C0
Ztau
Ztau.05
Cpm.crit
C0/Cpm.crit
mustar
sigstar
U
L
pL
pU
P.LU
```

1.2.1.3.1 Code 6.3

```
setwd("Your path here")
df1 <- read.csv("20240718 Chapter 6 RMSE Parameters.csv")
#
# this dataset has variables:
# tau = target value for X
# L = lower spec limit
# U = upper spec limit
# pL = Pr{X<L}
# pU = Pr{X>U}
# n = sample size for critical value
#
# It is assumed that the measurement, X, is normally distributed
#

attach(df1)


ZL <- qnorm(pL)# = (L - mustar)/sigstar
ZU <- qnorm(1-pU) # = (U - mustar)/sigstar
#
# Since pL and pU are given, ZL and ZU are also given.
# They are used to solve for mustar and sigstar.
# mustar and sigstar are the values of E[X] and SQRT(V[X])
# that yield:
# Pr{X<L} = pL and Pr{X>U} = pU exactly
#
# Note that the events {X<L} and {X>U} are mutually exclusive
# so Pr{X<L OR X>U} = pL + pU.
#
#
mustar <- (L*ZU - U*ZL) / (ZU - ZL)
sigstar <- (U - L) / (ZU - ZL)
Ztau <- (tau - mustar)/sigstar
lamda <- n*Ztau**2 #non-centrality parameter = 0 in
symmetric case
# By symmetric, we mean that L and U are symmetrically placed
around tau
# and pL = pU.
#
RMSEcrit <- sigstar * sqrt(qchisq(p=0.95,df=n,ncp=lamda))/sqrt(n)
#
# RMSEcrit is the 95th percentile of the sample RMSE if
# E[X] = mustar, V[X] = sigstar**2
```

```
#
ERMSE <- sqrt((mustar - tau)**2 + sigstar**2) #expected
value of RMSE
#
Ztau.alt <- seq(from=-4.0,to=4.0,by=0.01)
power <- 100*pchisq(q=n*(RMSEcrit/
sigstar)**2,df=n,ncp=n*Ztau.alt**2)
plot(x=Ztau.alt,y=power,main="Power Curve for RMSE",
     xlim=c(-5,5),ylim=c(0,100),xlab="Z-score for
tau",ylab="Pr{PASS}(%)")
abline(v=Ztau)
abline(h=5)
abline(h=95)
text(x=-3.3,y=97.5,labels="Pr{PASS}=95%")
text(x=-3.3,y=7.5,labels="Pr{PASS}=5%")
df2 <- cbind(tau,L,U,pL,pU,mustar,sigstar,RMSEcrit,ERMSE)

#
write.csv(df2,"20240718 RMSE Critical Values and Expected
Values.csv")
```

1.2.1.4 Chapter 8

1.2.1.4.1 Code 8.1

```
setwd("Your path here")
#
n.boot <- 2000
n.samp <- 100
estimate <- c()
scale.est <- c()
boot.samp <- c()
set.seed(13)
samp.shape <- 4.5
samp.scale <- 40
sample.data <- rgamma(n=n.samp,shape=samp.shape,scale=samp.scale)
#
for(i in 1:n.boot) {
  boot.samp <- sample(x=sample.data,size=n.samp,replace=TRUE)
  scale.est[i] <- mean(boot.samp*log(boot.samp)) - mean(boot.
samp)*mean(log(boot.samp))
  estimate[i] <- quantile(boot.samp,probs=0.75)
  rm(boot.samp)
  }
```

```
#
q.true <- qgamma(p=0.75,shape=samp.shape,scale=samp.scale)
q.interval <- quantile(estimate,probs=c(0.025,0.975),type=8)
q.LL <- q.interval[1]
q.UL <- q.interval[2]
q.med <- median(estimate)
scale.interval <- quantile(scale.est,probs=c(0.025,0.975),type=8)
scale.LL <- scale.interval[1]
scale.UL <- scale.interval[2]
scale.med <- median(scale.est)
q.true
q.LL
q.UL
q.med
#
samp.scale
scale.LL
scale.UL
scale.med

hist.boot <- hist(estimate,plot=FALSE)
hist.boot$density <- 100*hist.boot$counts / sum(hist.boot$counts)
plot(hist.boot,freq=FALSE,col="light
grey",xaxt="n",main="Histogram of Bootstrap 75th Percentile",
     xlab="75th Percentile",ylab="Percent",xlim=c(185,325),yli
m=c(0,50))
axis(side=1,at=seq(185,325,by=10))
abline(v=q.med,lty=1)
abline(v=q.true,lty=2)
abline(v=q.LL,lty=3)
abline(v=q.UL,lty=3)
legend(x=275,y=49,legend=c("Median","True Value","Conf.
Limit"),lty=c(1,2,3))
#
dev.new()
hist.scale <- hist(scale.est,plot=FALSE)
hist.scale$density <- 100*hist.scale$counts / sum(hist.
scale$counts)
plot(hist.scale,freq=FALSE,col="light
grey",xaxt="n",main="Histogram of Bootstrap Scale Parameter",
     xlab="Scale Parameter",ylab="Percent",xlim=c(15,70),yli
m=c(0,50))
axis(side=1,at=seq(15,70,by=5))
abline(v=scale.med,lty=1)
abline(v=samp.scale,lty=2)
```

```
abline(v=scale.LL,lty=3)
abline(v=scale.UL,lty=3)
legend(x=50,y=49,legend=c("Median","True Value","Conf.
Limit"),lty=c(1,2,3))
```

1.2.1.4.2 Code 8.2

```
setwd("Your path here")
piA <- 0.95
piB <- 0.85
n.reps <- 100
diff.p <- c()
OR <- c()
Odds.A <- c()
Odds.B <- c()
perm.A <- c()
perm.B <- c()
perm.diff <- c()
set.seed(26)
GroupA <- rbinom(n=n.reps,size=1,prob=piA)
GroupB <- rbinom(n=n.reps,size=1,prob=piB)
p.hat.A <- mean(GroupA)
p.hat.B <- mean(GroupB)
diff.obs <- p.hat.A - p.hat.B
Odds.A.obs <- p.hat.A / (1 - p.hat.A)
Odds.B.obs <- p.hat.B / (1 - p.hat.B)
OR.obs <- Odds.A.obs / Odds.B.obs
n.boot <- 2000
for(i in 1:n.boot){
  bootA <- sample(x=GroupA,size=n.reps,replace=TRUE)
  bootB <- sample(x=GroupB,size=n.reps,replace=TRUE)
  p.boot.A <- mean(bootA)
  p.boot.B <- mean(bootB)
  if(p.boot.A != 1) {
     Odds.A[i] <- p.boot.A / (1- p.boot.A)
     } else {
       Odds.A[i] <- 100
       }
  if(p.boot.B != 1) {
     Odds.B[i] <- p.boot.B / (1- p.boot.B)
     } else {
       Odds.B[i] <- 100
       }
#
```

```
  diff.p[i] <- p.boot.A - p.boot.B
  OR[i] <- Odds.A[i] / Odds.B[i]
  }
#
CI.diff <- quantile(diff.p,probs=c(0.025,0.975))
CI.OR <- quantile(OR,probs=c(0.025,0.975))
med.diff <- median(diff.p)
med.OR <- median(OR)

#
hist.diff <- hist(diff.p,plot=FALSE)
hist.diff$density <- 100*hist.diff$counts / sum(hist.diff$counts)
plot(hist.diff,freq=FALSE,col="light
grey",xaxt="n",main="Histogram
of PA - PB",xlab="PA - PB",ylab="Percentage")
# the xaxt="n" supresses the x-axis in the plot.
# The axis statement constructs the x-axis (side=1 is the x-axis)
axis(side=1,at=seq(from=0.0,to=.45,by=0.05))# this controls the
x-axis tick marks
abline(v=diff.obs,lty=1)
abline(v=CI.diff[1],lty=2)
abline(v=CI.diff[2],lty=2)
#
dev.new()
hist.OR <- hist(OR,plot=FALSE)
hist.diff$density <- 100*hist.diff$counts / sum(hist.diff$counts)
plot(hist.OR,freq=FALSE,col="light grey",xaxt="n",main="Histogram
of Odds.A / Odds.B",xlab="Odds Ratio",ylab="Percentage")
# the xaxt="n" supresses the x-axis in the plot.
# The axis statement constructs the x-axis (side=1 is the x-axis)
axis(side=1,at=seq(from=0.50,to=30,by=0.50))# this controls the
x-axis tick marks
abline(v=OR.obs,lty=1)
abline(v=CI.OR[1],lty=2)
abline(v=CI.OR[2],lty=2)
#
print("Proportions")
p.hat.A
p.hat.B
#
print("PA = PB")
diff.obs
CI.diff[1]
CI.diff[2]
#
```

```
print("OR.A / OR.B")
OR.obs
CI.OR[1]
CI.OR[2]
#
Permute.big <- as.numeric(rbind(GroupA,GroupB))
uni.sort <- runif(n=length(Permute.big),min=0,max=1)
Permute.sort <- data.frame(Permute.big,uni.sort)
Permute.sort.u <- Permute.sort[order(Permute.sort$uni.sort),]
binary.random <- Permute.sort.u$Permute.big
for(i in 1:n.boot){
  perm.A <- sample(x=binary.random,size=n.reps,replace=TRUE)
  perm.B <- sample(x=binary.random,size=n.reps,replace=TRUE)
  perm.diff[i] <- mean(perm.A) - mean(perm.B)
  }
F.perm <- ecdf(perm.diff)
if (diff.obs < 0) {
  p.value <- F.perm(diff.obs)
  } else {
    p.value <- 1- F.perm(diff.obs)
    }

print("Permutation Test for Difference")
p.value

dev.new()
hist.perm <- hist(perm.diff,plot=FALSE)
hist.perm$density <- 100*hist.perm$counts / sum(hist.perm$counts)
plot(hist.perm,freq=FALSE,col="light
grey",xaxt="n",main="Histogram
Permutation",xlab="PA - PB",ylab="Percentage")
# the xaxt="n" supresses the x-axis in the plot.
# The axis statement constructs the x-axis (side=1 is the x-axis)
axis(side=1,at=seq(from=-0.45,to=.45,by=0.05))# this controls the
x-axis tick marks
abline(v=diff.obs)
#
```

1.2.1.5 Chapter 9

1.2.1.5.1 Code 9.1

```
setwd("Your path here")
df1 <- read.csv("20240711 Chapter 9 Simple Linear Regression Data
```

```
Random.csv")
#
# Variables:
#
# X
# Y
#
attach(df1)
ols.line.pos <- lm(Y ~ X)
#ols.line.neg <- lm(Yneg ~ X)
intercept <- ols.line.pos$coefficients[1]
slope <- ols.line.pos$coefficients[2]
#
#int.neg <- ols.line.neg$coefficients[1]
#slope.neg <- ols.line.neg$coefficients[2]
#
plot(x=X,y=Y,main="Plot of Y vs. X")
#
dev.new()
plot(x=X,y=Y,main="Plot of Y vs. X with OLS Line",
     xlim=c(-1,70),ylim=c(0,100))
abline(a=intercept,b=slope)
#dev.new()
#plot(x=X,y=Yneg,main="Plot of Y vs. X Negative Slope",
#     xlim=c(0,70),ylim=c(-40,50))
#abline(a=int.neg,b=slope.neg)
```

1.2.1.5.2 Code 9.2

```
setwd("Your path here")
#
df1 <- read.csv("20240715 Chapter 9 Systolic with
Interaction.csv")
#
# Variables:
#
# Sodium
# Steps
# Systolic
#
attach(df1)
mult.1 <- lm(Systolic ~ Sodium + Steps)
summary(mult.1)
```

1.2.1.5.3 Code 9.3

```
setwd("Your path here")
df1 <- read.csv("20240815 Chapter 9 Collinearity.csv")
library(pls) #has function plsr
library(MASS) #has function lm.ridge
library(glmmLasso) #has function glmmLasso
#library(car) #needed to use function Anova() - not the same
as anova()
# Variables
#
# X.01
# X.02
# X.03
# X.04
# X.05
# X.06
# X.07
# X.08
# X.09
# X.10
# Response
attach(df1)
regressors <- as.matrix(cbind(const=1,X.01,X.02,X.03,X.04,X.05,
X.06,X.07,X.08,X.09,X.10))
model.pls <- plsr(Response ~.,data=df1,ncomp=8,method="simpls")
#model.pcr <- pcr(Response ~ X.01+X.02+X.03+X.04+X.05+X.06+X.07+
X.08+X.09+X.10,data=df1,ncomp=8,method="svdpc")
model.ridge <- lm.ridge(Response ~ X.01+X.02+X.03+X.04+X.05+X.06+
X.07+X.08+X.09+X.10,data=df1)
model.Lasso <- glmmLasso(fix=Response~X.01+X.02+X.03+X.04+X.05+X.0
6+X.07+X.08+X.09+X.10,rnd=NULL,lambda=1,data=df1)
plot(model.pls,ncomp=7,main="PLS,RIDGE,LASSO: Actual vs.
Predicted",xlab="Actual Response",pch=1,ylim=c(50,750))
#pred.pcr <- predict(model.pcr,ncomp=7)
pred.ridge <- regressors%*%coef(model.ridge)
pred.Lasso <- predict(model.Lasso)
#points(x=Response,y=pred.pcr,pch=2)
points(x=Response,y=pred.ridge,pch=3)
points(x=Response,y=pred.Lasso,pch=4)
legend(x=200,y=600,legend=c("PLS","RIDGE","LASSO"),pch=c(1,3,4))
```

1.2.1.6 Chapter 10

1.2.1.6.1 Code 10.1

```
setwd("Your path here")
#
df1 <- read.csv("20240718 Chapter 10 OD and Bacteria.csv")
#
# Variables:
# MEDIUM
# PREP
# OD
#
attach(df1)
OD.rank <- rank(OD,ties.method="average")
model.1 <- lm(OD ~ MEDIUM + PREP + MEDIUM:PREP)
summary(model.1)
tapply(OD,list(A=MEDIUM,B=PREP),FUN=mean)
tapply(OD,list(A=MEDIUM,B=PREP),FUN=sd)
interaction.plot(x.factor=MEDIUM,trace.factor=PREP,
                 response=OD,FUN=mean,main="Interaction of MEDIUM
and PREP")
wilcox.test(OD ~ factor(PREP))
#
anova.model <- lm(OD.rank ~ factor(MEDIUM) + factor(PREP) +
factor(MEDIUM):factor(PREP))
anova(anova.model)
```

1.2.1.6.2 Code 10.2

```
setwd("Your path here")
#
df1 <- read.csv("20240812 Chapter 10 Latin Square.csv")
#
# Variables:
# Lab
# Subject
# Group
# Order
# Meter.Type
# Meter.Reading
#
attach(df1)
RD <- 100*(Meter.Reading - Lab) / Lab
anova.model <- aov(RD ~ Group + factor(Subject)%in%Group +
factor(Order) + Meter.Type)
```

```
summary(anova.model)
tapply(RD,INDEX=Meter.Type,FUN=mean)
TukeyHSD(anova.model,which="Meter.Type")
```

1.2.1.7 Chapter 11

1.2.1.7.1 Code 11.1

```
setwd("Your path here")
#
n <- 52
alpha <- 0.05
lamda.0 <- 1
lamda.crit <- lamda.0 + qt(p=1-alpha,df=n)*sqrt(lamda.0/n)
t.crit <- sqrt(n)*(lamda.crit - lamda.0)/sqrt(lamda.0)
lamda.alt <- seq(from=lamda.0-0.5,to=lamda.0+2.0,by=0.005)
t.alt <- sqrt(n)*(lamda.crit - lamda.alt)/sqrt(lamda.alt)

power <- 100*pt(t.alt,df=n)#Pr{REJECT H0: Rate > lamda.0}
#
lamda.05 <- mean(lamda.alt[power>4 & power<6])
plot(x=lamda.alt,power,main="Power Curve for Poisson Rate
Parameter",
     xlim=c(lamda.0-0.5,2.05),ylim=c(0,100),xlab="Weekly
Rate",ylab="Power(%)")
abline(h=95)
abline(h=5)
abline(v=lamda.0,lty=2)
abline(v=lamda.05,lty=2)
text(x=0.7,y=7.5,labels="Power=5%")
text(x=0.7,y=93.0,labels="Power=95%")
```

1.2.1.8 Code 11.2

```
setwd("Your path here")
#
n <- 52
alpha <- 0.05
p.0 <- 0.05
tau.0 <- 1.0
lamda.0 <- -log(1-p.0)/tau.0
gamma.0 <- 1 / lamda.0
gamma.crit <- gamma.0*qchisq(p=alpha,df=2*n)/(2*n)
```

```
gamma.alt <- seq(from=0.001*gamma.0,to=1.5*gamma.0,by=0.1)
power <- 100*(1 - pchisq(q=2*n*gamma.crit/gamma.alt,df=2*n))
gamma.05 <- mean(gamma.alt[power>=4.0 & power <= 6.0])
gamma.001 <- mean(gamma.alt[power>=0.0 & power <= 0.5])
plot(x=gamma.alt,power,main="Power Curve for Exponential
Parameter",
     xlim=c(gamma.001,1.5*gamma.0),ylim=c(0,100),xlab="Expected
time-to-occurrence",
     ylab="Power(%)")
abline(v=gamma.05)
abline(h=5.0)
abline(h=95.0)
abline(v=gamma.05)
abline(v=gamma.0)
gamma.0.txt <- as.character(round(gamma.0,2))
gamma.05.txt <- as.character(round(gamma.05,2))
text(x=21,y=60,labels=gamma.0.txt)
text(x=10.5,y=60,labels=gamma.05.txt)
#
text(x=25,y=7.5,labels="Power=5%")
text(x=25,y=93.0,labels="Power=95%")
```

1.2.1.8.1 Code 11.3

```
setwd("Your path here")
#
df1 <- read.csv("20240723 Chapter 11 Time-to-Occurrence.csv")
#
# Variables:
#
# Index
# Time
# R.t
#
attach(df1)
H.t <- -log(R.t)
tsq <- Time**2
tcb <- Time**3
poly.1 <- lm(H.t ~ Time)
poly.2 <- lm(H.t ~ Time + tsq)
poly.3 <- lm(H.t ~ Time + tsq +tcb)
poly.2.noint <- lm(H.t ~ Time + tsq -1)
lamda <- mean(H.t)
summary(poly.1)
```

```
summary(poly.2)
summary(poly.3)
summary(poly.2.noint)
order.0 <- exp(-lamda*Time)
order.1 <- exp(-poly.1$coefficients[1] - poly.1$coefficients[2]*Time)
order.2
<- exp(-poly.2$coefficients[1] - poly.2$coefficients[2]*Time -
poly.2$coefficients[3]*tsq)
order.3
<- exp(-poly.3$coefficients[1] - poly.3$coefficients[2]*Time -
poly.3$coefficients[3]*tsq - poly.3$coefficients[4]*tcb)

plot(x=Time,y=R.t,main="Reliability/Survival
Curves",type="b",lty=1,pch=1,
     xlab="Time-to-Occurrence",ylab="Pr{Time>=t}")
points(x=Time,y=order.1,type="b",lty=2,pch=2)
points(x=Time,y=order.2,type="b",lty=3,pch=3)
points(x=Time,y=order.3,type="b",lty=4,pch=4)
points(x=Time,y=order.0,type="b",lty=5,pch=5)
legend(x=21,y=0.95,legend=c("Data","Linear","Quadratic","Cubic","
Exponential"),
       lty=c(1,2,3,4,5),pch=c(1,2,3,4,5))
df.out <- data.frame(cbind(Time,R.t,order.1,order.2,order.3))
write.csv(df.out,"20240723 Reliability Estimates with Polynomial
Fits.csv")
```

1.2.1.8.2 Code 11.3

```
setwd("Your path here")
df1 <- read.csv("20240724 accelerated life test.csv")

attach(df1)
# inputs:
# Test = index
# TempC = test temp in oC
# Time = test time in hours
# n = number of units on test
# s = number surviving units after test
#
# the R function "log" is the natural logarithm
#
TK1 <- TempC[1] + 273
TK2 <- TempC[2] + 273
TK0 <- 25 + 273 #Room Temperature
```

```
n1 <- n[1]
n2 <- n[2]
s1 <- s[1]
s2 <- s[2]
lnA <- (TK2)/(TK2 - TK1)*(log(-log(s2/n2)) + TK1*log(-log(s1/
n1))/TK2 + log(Time))
B <- TK1*(lnA - log(-log(s1/n1)) + log(Time))
K <- exp(-B/TK2) / exp(-B/TK1)
K.0 <- 26 #desired acceleration factor
Pa <- B*TK0 / (B - TK0*log(K.0))
TOC <- Pa - 273
df2 <- cbind(lnA,B,K,K.0,Pa,TOC)
write.csv(df2,file="20240724 accelerated life test output.csv")
```

1.2.1.9 Chapter 12

1.2.1.9.1 Code 12.1

```
setwd("Your path here")
df1 <- read.csv("20240725 Chapter 12 Full Factorial
Experiment.csv")
#
# Variables
#
# Run
# Temp oC
# Pres kP
# Time min
# Yield Liters
#
library(olsrr)
attach(df1)
fTemp <- factor(Temp)
fPres <- factor(Pres)
fTime <- factor(Time)
model.full <- lm(Yield ~ fTemp + fPres + fTime +
                 fTemp:fPres + fTemp:fTime + fPres:fTime +
                 fTemp:fPres:fTime)
#
anova(model.full)
#
model.reg <- lm(Yield ~ Temp + Pres + Time +
                Temp:Pres + Temp:Time + Pres:Time +
                Temp:Pres:Time)
```

```
#
model.red <- lm(Yield ~ Temp + Pres +
                 Temp:Pres)
#
summary(model.reg)
#
pred.reg <- as.numeric(predict(model.reg))
pred.line <- lm(pred.reg ~ Yield)
int <- pred.line$coefficients[1]
slope <- pred.line$coefficients[2]

plot(x=Yield,y=pred.reg,main="Predicted vs. Observed Yield",
     xlab="Observed Yield (L)",ylab="Predicted Yield (L)")
abline(a=int,b=slope)
ols_mallows_cp(model.reg,fullmodel=model.reg)
#
summary(model.red)
#
pred.red <- as.numeric(predict(model.red))
pred.line.red <- lm(pred.red ~ Yield)
int.r <- pred.line.red$coefficients[1]
slope.r <- pred.line.red$coefficients[2]
dev.new()
plot(x=Yield,y=pred.red,main="Predicted vs. Observed Yield -
Reduced Model",
     xlab="Observed Yield (L)",ylab="Predicted Yield (L) -
Reduced Model")
abline(a=int.r,b=slope.r)
ols_mallows_cp(model.red,fullmodel=model.reg)
#
model.min <- lm(Yield ~ Temp:Pres)
#
summary(model.min)
ols_mallows_cp(model.min,fullmodel=model.reg)
```

1.2.1.9.2 Code 12.2

```
setwd("Your path here")
df1 <- read.csv("20240726 Chapter 12 Half-Fraction 5-Factor Yield
Experiment.csv")
#
# Variables
#
# Rep
```

```
# Run
# Temp oC
# Pres kP
# Time min
# Speed
# pH
# Yield Liters
#
library(olsrr)
attach(df1)
fTemp <- factor(Temp)
fPres <- factor(Pres)
fTime <- factor(Time)
fSpeed <- factor(Speed)
fpH <- factor(pH)
model.full <- lm(Yield ~ fTemp + fPres + fTime +
                  fSpeed + fpH +
                  fTemp:fPres + fTemp:fTime + fPres:fTime +
                  fTemp:fSpeed + fTemp:fpH +
                  fPres:fSpeed + fPres:fpH +
                  fSpeed:fpH)
anova(model.full)
#
model.reg <- lm(Yield ~ Temp + Pres + Time +
                  Speed + pH +
                  Temp:Pres + Temp:Time + Pres:Time +
                  Temp:Speed + Temp:pH +
                  Pres:Speed + Pres:pH +
                  Speed:pH)
#
summary(model.reg)
#
pred.reg <- as.numeric(predict(model.reg))
pred.line <- lm(pred.reg ~ Yield)
int <- pred.line$coefficients[1]
slope <- pred.line$coefficients[2]
plot(x=Yield,y=pred.reg,main="Predicted vs. Observed Yield",
     xlab="Observed Yield (L)",ylab="Predicted Yield (L)")
abline(a=int,b=slope)
ols_mallows_cp(model.reg,fullmodel=model.reg)
#
#
```

1.2.1.10 Chapter 13

1.2.1.10.1 Code 13.1

```
setwd("Your path here")
#
df1 <- read.csv("20240729 Chapter 13 OD and Bacteria Redux.csv")
#
# Variables:
# MEDIUM
# PREP
# OD
#
attach(df1)
OD.rank <- rank(OD,ties.method="average")
model.1 <- lm(OD ~ factor(MEDIUM) + factor(PREP) +
factor(MEDIUM):factor(PREP))
anova(model.1)
tapply(OD,list(A=MEDIUM,B=PREP),FUN=mean)
tapply(OD,list(A=MEDIUM,B=PREP),FUN=sd)
interaction.plot(x.factor=MEDIUM,trace.factor=PREP,
                 response=OD,fun=mean,main="Interaction of MEDIUM
and PREP")
wilcox.test(OD ~ factor(MEDIUM))
kruskal.test(OD ~ PREP)
#
anova.model <- lm(OD.rank ~ factor(MEDIUM) + factor(PREP) +
factor(MEDIUM):factor(PREP))
anova(anova.model)
dev.new()
interaction.plot(x.factor=MEDIUM,trace.factor=PREP,
                 response=OD.rank,fun=mean,main="Interaction of
MEDIUM and PREP Ranks")
```

1.2.1.10.2 Code 13.2

```
setwd("Your path here ")
#
data.mat <- matrix(c(4,26,36,34),nrow=2,
                   dimnames=list(Gender=c("Men","Women"),Position
=c("Mgr","Non-Mgr")))
data.mat
test.stat <- fisher.test(x=data.mat)
test.stat
```

1.2.1.10.3 Code 13.3

```
setwd("Your path here ")
#
#  Power for an r x c contingency table using a Chi-Squared
statistic
#  n.size is the number of individuales having the row and column
variables
#  tallied.
#
#  H0: rows and columns are uncorrelated (surrogate for
independence)
#  Ha: results in rows and columns depend upon each other.
#
#
prob.alt <- c()
Non.Cent <- c()
power <- c()
n.size <- 100
n.txt <- as.character(round(n.size,0))
n.rows <- 2
n.cols <- 2
alpha <- 0.05
d.f. <- (n.rows-1)*(n.cols-1)
ChiSq.crit <- qchisq(p=1-alpha,df=d.f.)
ChiSq.crit
prob.0 <- pchisq(q=ChiSq.crit,df=d.f.)
prob.0 #just a check to make sure ChiSq.crit is correct
#
# correl is also called the Kaplan "effect";  correl = 0 means
independence
# of rows and columns in an r x c contingency table
#
correl <- seq(from=0.0,to=0.99,by=0.01)
Non.Cent <- n.size*correl**2
power <- 1-pchisq(q=ChiSq.crit,df=d.f.,ncp=Non.Cent)
#
#  The higher the sample size (n.size) the lower c95 will be
#
c95 <- mean(correl[power >= 0.94 & power <= 0.96])
c95.txt <- as.character(round(c95,2))
plot(x=correl,y=100*power,main="Power for Chi-Squared Test of
Independence",
     sub="H0: Correlation = 0",
     type="b",xlab="Correlation between Rows and Columns",
     ylab="Power=Pr{Reject Independence}")
abline(h=5)
```

```
abline(h=95)
abline(v=c95)
text(x=0.8,y=7.5,labels="Pr{REJECT H0 = 5%}")
text(x=0.8,y=92.5,labels="Pr{REJECT H0 = 95%}")
text(x=0.725,y=20,labels="n = ")
text(x=0.8,y=20,labels=n.txt)
text(x=c95+0.035,y=60,labels=c95.txt)
```

1.2.1.11 Chapter 14

1.2.1.11.1 Code 14.1

```
setwd("Your path here")
#
df1 <- read.csv("20240731 Chapter 14 Logistic and Continuous Data
with 10 Regressors.csv")
#
# Variables:
# X.01-X.10
# Response
# Response.B
#
#

attach(df1)
logistic.model <- glm(Response.B ~ X.01 + X.02 + X.03 + X.04
+ X.05 +
                                          X.06 + X.07 + X.08 +
X.09 + X.10,
                       data=df1,family=binomial(link="logit"))
logistic.predict <- 1 / (1 + exp(-predict(logistic.model,df1)))
summary(logistic.model)
plot(x=X.04,y=logistic.predict)
```

1.2.1.11.2 Code 14.2

```
setwd("Your path here")
#
df1 <- read.csv("20240731 Chapter 14 Logistic Regression
ANOVA.csv")
#
# Variables:
# Robot
```

```
# Product
# test
#
#
attach(df1)
logistic.model <- glm(test ~ factor(Robot) + factor(Product) +
factor(Robot):factor(Product),
                     family=binomial(link="logit"))
logistic.predict <- 1 / (1 + exp(-predict(logistic.model,df1)))
summary(logistic.model)
plot(x=Robot,y=logistic.predict)
dev.new()
interaction.plot(x.factor=Robot,response=logistic.
predict,fun=mean,trace.factor=Product,
                main="Interaction between Robot and Product -
Pr{Correct  Placement}")
```

1.2.1.11.3 Code 14.3

```
setwd("Your path here")
#
df1 <- read.csv("20240731 Chapter 14 Nonlinear Model.csv")
#
# Variables:
#
# U
# X
# Y
attach(df1)
nls.model <- nls(Y ~ exp(B0*X) / (1 + exp(B1*X)),start=list(B0=
0.5,B1=1),
                control=list(maxiter=100))
nls.pred <- as.numeric(predict(nls.model))

plot(x=X,y=Y,main="Scatter Plot of Nonlinear Data")
parms <- summary(nls.model)
nls.test <- exp(parms$coefficients[1]*X) / (1 +
exp(parms$coefficients[2]*X))
LogY <- log(Y)
lin.mod <- lm(Y ~ X)
lin.pred <- as.numeric(predict(lin.mod))
exp.pred <- exp(-lin.pred)
dev.new()
plot(x=X,y=Y,main="Nonlinear Data with Curve Fit",type="p",pch=1,
```

```
ylim=c(0.5,1.05))
points(x=X,y=lin.pred,pch=3,type="p",lty=2)
points(x=X,y=nls.test,pch=2,type="p",lty=3)
legend(x=4,y=0.70,legend=c("Data","Nonlinear Fit","Linear
Fit"),pch=c(1,2,3))
```

1.2.1.12 Chapter 15

1.2.1.12.1 Code 15.1

```
setwd("Your path here")
#
df1 <- read.csv("20240731 Chapter 14 Logistic and Continuous Data
with 10 Regressors.csv")
#
# Variables:
# X.01-X.10
# Response
# Response.B
#
#

attach(df1)
#
big.model <- lm(Response ~ X.01+X.02+X.03+X.04+X.05+X.06+X.07+X.0
8+X.09+X.10+
                           X.01:X.03 + X.01:X.06 + X.01:X.07 +
X.01:X.08 + X.01:X.09 + X.01:X.10 +
                           X.03:X.06 + X.03:X.07 + X.03:X.08 +
X.03:X.09 + X.03:X.10 +
                           X.06:X.07 + X.06:X.08 + X.06:X.09 +
X.06:X.10 +
                           X.07:X.08 + X.07:X.09 + X.07:X.10 +
                           X.08:X.09 + X.08:X.10 +
                           X.09:X.10)
summary(big.model)

#step.forward <- step(big.model,direction="forward",
#                     scope=list(lower=Response~X.01,upper=Respon
se~X.01+X.02+X.03+X.04+X.05+X.06+X.07+X.08+X.09+X.10))
#
step.backward <- step(big.model,direction="backward")
fitted.backward <- as.numeric(step.backward$fitted.values)
fitted.observed <- lm(fitted.backward ~ Response)
```

```
int <- fitted.observed$coefficients[1]
slo <- fitted.observed$coefficients[2]
plot(x=Response,y=fitted.backward,main="Predicted vs. Observed
Response: Final Stepwise Model",
     xlab="Observed Response",ylab="Predicted Response")
abline(a=int,b=slo)
#
logistic.model <- glm(Response.B ~ X.01 + X.02 + X.03 + X.04
+ X.05 +
                                   X.06 + X.07 + X.08 + X.09
+ X.10 +
                                   X.01:X.03 + X.01:X.06 +
X.01:X.07 + X.01:X.08 + X.01:X.09 + X.01:X.10 +
                                   X.03:X.06 + X.03:X.07 +
X.03:X.08 + X.03:X.09 + X.03:X.10 +
                                   X.06:X.07 + X.06:X.08 +
X.06:X.09 + X.06:X.10 +
                                   X.07:X.08 + X.07:X.09 +
X.07:X.10 +
                                   X.08:X.09 + X.08:X.10 +
                                   X.09:X.10,data=df1,family=bino
mial(link="logit"))
#
summary(logistic.model)
#
step.logistic <- step(logistic.model,direction="backward")
logistic.reduced <- glm(Response.B ~ X.01 + X.02 + X.04 + X.05 +
                                   X.06 + X.07 + X.08 + X.09
+ X.10 +
                                   X.01:X.07 +
                                   X.03:X.09 +
                                   X.06:X.07 + X.06:X.08 +
                                   X.07:X.10 +
                                   X.08:X.09 + X.08:X.10,
                                   data=df1,family=binomial(link
="logit"))

#
step.log.pred <- as.numeric(step.logistic$fitted.values)
reduced.log.pred <- logistic.reduced$fitted.values
prob.reg <- lm(reduced.log.pred ~ step.log.pred)
int.prob <- prob.reg$coefficients[1]
slo.prob <- prob.reg$coefficients[2]
dev.new()
plot(x=step.log.pred,y=reduced.log.pred,main="Predcited
```

```
Probabilities: Reduced vs. Final Stepwise Models",
     xlab="Stepwise Model Predictions",ylab="Reduced Model
Predictions")
abline(a=int.prob,b=slo.prob)
#
logistic.further <- glm(Response.B ~ X.01 + X.02 + X.04 +
                                   X.06 + X.07 + X.08 + X.09
+ X.10 +

                                   X.01:X.07 +
                                   X.03:X.09 +
                                   X.06:X.07 +
                                   X.07:X.10 +
                                   X.08:X.09 + X.08:X.10,
                                   data=df1,family=binomial(lin
k="logit")
```

1.2.1.12.2 Code 15.2

```
setwd("Your path here")
#
df1 <- read.csv("20240731 Chapter 14 Logistic and Continuous Data
with 10 Regressors.csv")
#
# Variables:
# X.01-X.10
# Response
# Response.B
#
#
library(BMA)
library(glmulti)
attach(df1)
#
bma.model <- bic.glm(f=Response ~ X.01+X.02+X.03+X.04+X.05+X.06+
X.07+X.08+X.09+X.10+
                           X.01:X.03 + X.01:X.06 + X.01:X.07 +
X.01:X.08 + X.01:X.09 + X.01:X.10 +
                           X.03:X.06 + X.03:X.07 + X.03:X.08 +
X.03:X.09 + X.03:X.10 +
                           X.06:X.07 + X.06:X.08 + X.06:X.09 +
X.06:X.10 +
                           X.07:X.08 + X.07:X.09 + X.07:X.10 +
                           X.08:X.09 + X.08:X.10 +
```

```
                           X.09:X.10,data=df1,
                           glm.family=gaussian(link="identity"),o
ccam.window=TRUE)
#
bma.predicted <- as.numeric(predict(bma.model,newdata=df1))
#
glmulti.model <- glmulti(Response ~ X.01+X.02+X.03+X.04+X.05+X.06
+X.07+X.08+X.09+X.10,
                         level=2,method="g",maxsize=20,crit="aic"
,plotty=FALSE)
#
glmulti.predicted <- predict(glmulti.model,newdata=df1)
#
big.model <- lm(Response ~ X.01+X.02+X.03+X.04+X.05+X.06+X.07+X.0
8+X.09+X.10+
                           X.01:X.03 + X.01:X.06 + X.01:X.07 +
X.01:X.08 + X.01:X.09 + X.01:X.10 +
                           X.03:X.06 + X.03:X.07 + X.03:X.08 +
X.03:X.09 + X.03:X.10 +
                           X.06:X.07 + X.06:X.08 + X.06:X.09 +
X.06:X.10 +
                           X.07:X.08 + X.07:X.09 + X.07:X.10 +
                           X.08:X.09 + X.08:X.10 +
                           X.09:X.10)
summary(big.model)
big.predicted <- as.numeric(big.model$fitted.values)

#step.forward <- step(big.model,direction="forward",
#                       scope=list(lower=Response~X.01,upper=Respon
se~X.01+X.02+X.03+X.04+X.05+X.06+X.07+X.08+X.09+X.10))
#
step.backward <- step(big.model,direction="backward")
fitted.backward <- as.numeric(step.backward$fitted.values)
fitted.observed <- lm(fitted.backward ~ Response)
int <- fitted.observed$coefficients[1]
slo <- fitted.observed$coefficients[2]
plot(x=Response,y=fitted.backward,main="Predicted vs. Observed
Response: Final Stepwise Model",
     xlab="Observed Response",ylab="Predicted Response")
abline(a=int,b=slo)
#
logistic.model <- glm(Response.B ~ X.01 + X.02 + X.03 + X.04
+ X.05 +
                                   X.06 + X.07 + X.08 + X.09
+ X.10 +
```

```
                                      X.01:X.03 + X.01:X.06 +
X.01:X.07 + X.01:X.08 + X.01:X.09 + X.01:X.10 +
                                      X.03:X.06 + X.03:X.07 +
X.03:X.08 + X.03:X.09 + X.03:X.10 +
                                      X.06:X.07 + X.06:X.08 +
X.06:X.09 + X.06:X.10 +
                                      X.07:X.08 + X.07:X.09 +
X.07:X.10 +
                                      X.08:X.09 + X.08:X.10 +
                                      X.09:X.10,data=df1,family=bino
mial(link="logit"))
#
summary(logistic.model)
#
step.logistic <- step(logistic.model,direction="backward")
logistic.reduced <- glm(Response.B ~ X.01 + X.02 + X.04 + X.05 +
                                      X.06 + X.07 + X.08 + X.09
+ X.10 +
                                      X.01:X.07 +
                                      X.03:X.09 +
                                      X.06:X.07 + X.06:X.08 +
                                      X.07:X.10 +
                                      X.08:X.09 + X.08:X.10,
                                      data=df1,family=binomial(link
="logit"))

#
step.log.pred <- as.numeric(step.logistic$fitted.values)
reduced.log.pred <- logistic.reduced$fitted.values
prob.reg <- lm(reduced.log.pred ~ step.log.pred)
int.prob <- prob.reg$coefficients[1]
slo.prob <- prob.reg$coefficients[2]
dev.new()
plot(x=step.log.pred,y=reduced.log.pred,main="Predcited
Probabilities: Reduced vs. Final Stepwise Models",
     xlab="Stepwise Model Predictions",ylab="Reduced Model
Predictions")
abline(a=int.prob,b=slo.prob)
#
logistic.further <- glm(Response.B ~ X.01 + X.02 + X.04 +
                                      X.06 + X.07 + X.08 + X.09
+ X.10 +
                                      X.01:X.07 +
                                      X.03:X.09 +
                                      X.06:X.07 +
```

```
                                          X.07:X.10 +
                                          X.08:X.09 + X.08:X.10,
                                          data=df1,family=binomial(link
="logit"))
#
dev.new()
plot(x=Response,y=bma.predicted,main="BMA, Stepwise, Full Linear,
GLMULTI vs. Observed Response",pch=1,
     ylab="Predicted Response",xlab="Observed Response")
points(x=Response,y=fitted.backward,pch=2)
points(x=Response,y=big.predicted,pch=3)
points(x=Response,y=glmulti.predicted$averages,pch=4)
legend(x=0,y=170,legend=c("BMA","Stepwise","Full Linear","GLMULTI
"),pch=c(1,2,3,4))
```

1.2.1.13 Chapter 16

1.2.1.13.1 Code 16.1

```
setwd("Your path here")
df1 <- read.csv("20240805 Chapter 16 Multivariate Data with
Groups New.csv")

attach(df1)
# the dataset has 11 columns:
# Unit, Group, X0-X9
library(MASS)
n.clusters <- 4
#input.data <- data.frame(cbind(X0,X1,X2,X3,X4,X5,X6,X7,X8,X9,X10
,X11,X12,X13,X14,X15))
input.data <- data.frame(cbind(X0,X1,X3,X4,X5,X6,X7,X8,X9))
alt.input <- data.frame(cbind(X0,X1,X2,X5,X6,X7,X8,X9))
clusk <- kmeans(input.data,centers=n.clusters)
cluster.id <- clusk$cluster
p.comps <- princomp(input.data,cor=TRUE,scores=TRUE)
var.pcs <- p.comps$sdev**2
total.var <- sum(var.pcs)
pct.var <- 100*var.pcs / total.var
pct.var
#
#df.df1 <- data.frame(df1)
linear.disc <- lda(df1$Group ~X0+X1+X2+X3+X4+X5+X6+X7+X8+X9,data=
df1,CV=FALSE)
#linear.disc <- lda(df1$Group ~X0+X1+X2+X5+X6+X7+X8+X9,df1
```

```
,CV=TRUE)
linear.disc
lda.pred <- predict(linear.disc)
dev.new()
ldahist(data=lda.pred$x[,1],g=df1$Group,nbin=12)
#dev.new()
#ldahist(data=lda.pred$x[,2],g=df1$Group,nbin=12)
#dev.new()
#ldahist(data=lda.pred$x[,1],g=df1$Group,nbin=12)
#
data.w.pcomps <- cbind(input.data,df1$Group,cluster.id,p.
comps$scores[,1])
detach(df1)
attach(data.w.pcomps)
pc1.1 <- Comp.1[which(cluster.id==1)]
pc2.1 <- Comp.2[which(cluster.id==1)]
pc1.2 <- Comp.1[which(cluster.id==2)]
pc2.2 <- Comp.2[which(cluster.id==2)]
pc1.3 <- Comp.1[which(cluster.id==3)]
pc2.3 <- Comp.2[which(cluster.id==3)]
pc1.4 <- Comp.1[which(cluster.id==4)]
pc2.4 <- Comp.2[which(cluster.id==4)]
#
dev.new()
plot(pc1.1,pc2.1,main="Cluster Assignment",pch=1,xlim=c(-3.5,10),
ylim=c(-3,3),xlab="Prin.Comp. 1",ylab="Prin.Comp 2")
points(pc1.2,pc2.2,pch=2)
points(pc1.3,pc2.3,pch=3)
points(pc1.4,pc2.4,pch=4)
#legend(x=6,y=-1,legend=c("Cluster 1","Cluster 2","Cluster
3"),pch=c(1,2,3))
legend(x=6,y=-1,legend=c("Cluster 1","Cluster 2","Cluster
3","Cluster 4"),pch=c(1,2,3,4))
#

agreement <- cbind(df1$Group,cluster.id)
cross.tab <- table(df1$Group,cluster.id)
cross.tab
pc.1 <- p.comps$scores[,1]
pc.2 <- p.comps$scores[,2]
fGroup <- factor(df1$Group)
fclust <- factor(cluster.id)
mod.group <- lm(pc.1 ~ fGroup)
anova.group <- aov(mod.group)
mod.clust <- lm(pc.1 ~ fclust)
```

```
anova.clust <- aov(mod.clust)
HSD.group <- TukeyHSD(anova.group,which="Group",conf.level=0.95)
HSD.clust <- TukeyHSD(anova.clust,which="clust",conf.level=0.95)
HSD.group
HSD.clust
```

1.2.1.14 Chapter 17

1.2.1.14.1 Code 17.1

```
options(na.action=na.exclude)
#
# setwd tells R which folder to use for this script
#

setwd("Your path here")
#
# This is an MCMC for a beta-binomial conjugate pair
# The conjugacy is used to illustrate how MCMC yields
approximately
# the correct posterior distribution
#
set.seed(65)
pstore <- c()
#
a.beta <- 1
b.beta <- .25
alphapri <- a.beta
betapri <- b.beta
#
#  Values of the parameter will be randomly generated
#  from the Beta prior, to illustrate the shape of
#  the prior density.
#
beta.prior <- rbeta(n=2000,shape1=a.beta,shape2=b.beta)
beta.hist <- hist(beta.prior,plot=FALSE)
beta.hist$density <- 100*beta.hist$counts / sum(beta.hist$counts)
plot(beta.hist,freq=FALSE,col="light grey",main="Prior
Distribution for Binomial Parameter",xaxt="n",yaxt="n",
     xlim=c(0,1),ylim=c(0,max(beta.
hist$density)+1),xlab="Binomial Parameter",ylab="Frequency (%)")
axis(side=1,at=seq(from=0,to=1,by=0.05))
axis(side=2,at=seq(from=0,to=100,by=5))
#
```

```
#  Now create the observed data:
#
sampsize <- 50
y <- 40
phat <- y / sampsize
p0 <- a.beta / (a.beta + b.beta)
sig.beta <- sqrt((a.beta*b.beta) /(((a.beta+b.beta)**2)*(a.
beta+b.beta+1)))
a.post <- a.beta + y
b.post <- b.beta + sampsize - y
nmcmc <- 20
for (i in 1:nmcmc) {
  p1 <- min(abs(rnorm(n=1,mean=p0,sd=0.05)),1)#use folded &
truncated normal to generate p1
#                                                  insures p1 will
be in interval (0,1)
#  p1 <- runif(n=1,min=0,max=1)
  num <- dbeta(x=p1,shape1=alphapri,shape2=betapri)*dbinom(x=y,si
ze=sampsize,prob=p1)
  den <- dbeta(x=p0,shape1=alphapri,shape2=betapri)*dbinom(x=y,si
ze=sampsize,prob=p0)
  rat <- num / den
  uvar <- runif(n=1,min=0,max=1)
  if (rat >= uvar) {
    pstore[i] <- p1
    } else {
    pstore[i] <- p0
    }
  p0 <- pstore[i] #change p0 to current value
  }
#
#
#
alphapos <- alphapri + sampsize
betapos <- betapri + sampsize - y
meanpos <- alphapos / (alphapos + betapos)
dev.new()
prob.hist <- hist(pstore,plot=FALSE)
prob.hist$density <- 100*prob.hist$counts / sum(prob.hist$counts)
#
LL <- min(pstore) - 0.01
UL <- max(pstore) + 0.01
p.list <- seq(from=0.60,to=1,by=0.01)
p.dens <- dbeta(x=p.list,shape1=a.post,shape2=b.post)
plot(prob.hist,freq=FALSE,main="Histogram of Posterior
```

```
Sample",col="light grey",
      xaxt="n",yaxt="n",xlim=c(0.60,1.0),ylim=c(0,max(prob.
hist$density)+1),xlab="Binomial Parameter",ylab="Frequency (%)")
axis(side=1,at=seq(from=0.60,to=1.0,by=0.01))#x-axis
axis(side=2,at=seq(from=0,to=max(prob.
hist$density)+1,by=1))#y-axis
#points(x=p.list,y=p.dens,type="l",lty=1)
lowlim <- quantile(x=pstore,probs=c(0.025),type=8)
medprob <- quantile(x=pstore,probs=c(0.50),type=8)
upplim <- quantile(x=pstore,probs=c(0.975),type=8)
lowlim
medprob
upplim
conjugate.low <- qbeta(p=0.025,shape1=a.post,shape2=b.post)
conjugate.med <- qbeta(p=0.500,shape1=a.post,shape2=b.post)
conjugate.upp <- qbeta(p=0.975,shape1=a.post,shape2=b.post)
conjugate.low
conjugate.med
conjugate.upp

df2 <- cbind(alphapri,betapri,alphapos,betapos,sampsize,y,lowlim,
upplim,pstore)
write.csv(df2,"20240805 Chapter 17 Beta-Binomial posterior.csv")
```

1.2.1.14.2 Code 17.2

```
options(na.action=na.exclude)
setwd("Your path here")
#
# This is an MCMC for a normal-normal conjugate pair
# The conjugacy is used to illustrate how MCMC yields
approximately
# the correct posterior distribution
#
set.seed(26)
mustore <- c()
#
mupri <- 90
taupri <- 5
#
#  Values of the parameter will be randomly generated
#  from the Normal prior, to illustrate the shape of
#  the prior density.
#
```

```
#  Now create the observed data:
#
sampsize <- 50
n <- sampsize
xbar <- 93
sigdat <- 10
p0 <- mupri
mu.post <- ((sigdat**2)*mupri + n*(taupri**2)*xbar) /
(n*taupri**2 + sigdat**2)
tau.post <- sqrt((taupri*sigdat)**2 / (n*taupri**2 + sigdat**2))
nmcmc <- 2000
for (i in 1:nmcmc) {
  p1 <- rnorm(n=1,mean=p0,sd=taupri)
  num <- dnorm(x=p1,mean=p0,sd=taupri)*dnorm(x=xbar,mean=p1,sd=si
gdat/sqrt(n))
  den <- dnorm(x=p0,mean=p1,sd=taupri)*dnorm(x=xbar,mean=p0,sd=si
gdat/sqrt(n))
  rat <- num / den
  uvar <- runif(n=1,min=0,max=1)
  if (rat >= uvar) {
    mustore[i] <- p1
    }
  else {
    mustore[i] <- p0
    }
  p0 <- mustore[i] #change p0 to current value
  }
#
#
#
dev.new()
prob.hist <- hist(mustore,plot=FALSE)
prob.hist$density <- 100*prob.hist$counts / sum(prob.hist$counts)
#
LL <- round(mu.post-6*tau.post,2)
UL <- round(mu.post+6*tau.post,2)
mu.norm <- seq(from=LL,to=UL,by=0.25)
post.dens <- 100*dnorm(x=mu.norm,mean=mu.post,sd=tau.post)
plot(prob.hist,freq=FALSE,main="Histogram of Posterior
Sample",col="light grey",
      xaxt="n",yaxt="n",xlim=c(mu.post-4*tau.post,mu.post+4*tau.
post),ylim=c(0,max(prob.hist$density)+1),
      xlab="Mu",ylab="Frequency (%)")
axis(side=1,at=seq(from=LL,to=UL,by=0.25))#x-axis
axis(side=2,at=seq(from=0,to=max(prob.
```

```
hist$density)+1,by=1))#y-axis
points(x=mu.norm,y=post.dens,type="l",lty=1)
lowlim <- quantile(x=mustore,probs=c(0.025),type=8)
medprob <- quantile(x=mustore,probs=c(0.50),type=8)
upplim <- quantile(x=mustore,probs=c(0.975),type=8)
lowlim
medprob
upplim
conjugate.low <- qnorm(p=0.025,mean=mu.post,sd=tau.post)
conjugate.med <- qnorm(p=0.500,mean=mu.post,sd=tau.post)
conjugate.upp <- qnorm(p=0.975,mean=mu.post,sd=tau.post)
conjugate.low
conjugate.med
conjugate.upp

df2 <- cbind(mupri,taupri,mu.post,tau.post,n,xbar,lowlim,uppli
m,mustore)
write.csv(df2,"20240805 Chapter 17 Normal-Normal posterior.csv")
```

1.2.1.15 Chapter 18

1.2.1.15.1 Code 18.1

```
setwd("Your path here")
#
df1 <- read.csv("20240806 Chapter 18 Neural Network Data.csv")
#
# Variables:
# X0-X9, Unit, Group, Response, Train.Data
#
#
#
#
library(nnet)
attach(df1)
train.set <- subset(df1,subset=Train.Data==1)
test.set <- subset(df1,subset=Train.Data==0)
detach(df1)
attach(train.set)
big.net <- nnet(Response~X0+X1+X2+X3+X4+X5+X6+X7+X8+X9,size=4,dat
a=train.set,skip=TRUE,linout=TRUE)
#
big.pred <- big.net$fitted.values
linear.model <- lm(big.pred~Response)
```

```
int <- linear.model$coefficients[1]
slo <- linear.model$coefficients[2]
plot(x=train.set$Response,y=big.pred,main="Neural Network
Predictions vs. Observed Response",
     xlab="Observed",ylab="Predicted")
abline(a=int,b=slo)
detach(trian.set)
attach(test.set)
test.pred <- predict(big.net,newdata=test.set,type="raw")
linear.oos <- lm(test.pred~Response)
int.oos <- linear.oos$coefficients[1]
slo.oos <- linear.oos$coefficients[2]
dev.new()
plot(x=Response,y=test.pred,main="Neural Network Out-of-Sample
Predictions vs. Observed Responses",
     xlab="Observed",ylab="Predicted")
abline(a=int.oos,b=slo.oos)
```

1.2.1.15.2 Code 18.2

```
setwd("Your path here")
#
df1 <- read.csv("20240806 Chapter 18 Neural Network Data.csv")
#
# Variables:
# X0-X9, Unit, Group, Response, Train.Data
#
#
#
#
library(nnet)
library(rpart)
attach(df1)
train.set <- subset(df1,subset=Train.Data==1)
test.set <- subset(df1,subset=Train.Data==0)
detach(df1)
attach(train.set)
big.net <- nnet(Response~X0+X1+X2+X3+X4+X5+X6+X7+X8+X9,size=4,dat
a=train.set,skip=TRUE,linout=TRUE)
net.pred <- predict(big.net,newdata=train.set)
big.cart <- rpart(Response~X0+X1+X2+X3+X4+X5+X6+X7+X8+X9,data=
train.set)
#
rsq.rpart(big.cart)
```

```
#
dev.new()
plot(big.cart,xlim=c(-0.1,12),ylim=c(-0.1,1.15),main="CART
Regression Tree")
text(big.cart,use.n=TRUE)
cart.pred <- predict( big.cart,newdata=train.set)
dev.new()
linear.model <- lm(cart.pred~train.set$Response)
int <- linear.model$coefficients[1]
slo <- linear.model$coefficients[2]
plot(x=train.set$Response,y=cart.pred,main="CART Predictions vs.
Observed Response",
     xlab="Observed",ylab="Predicted")
abline(a=int,b=slo)
#
detach(train.set)
attach(test.set)
net.pred.test <- predict(big.net,newdata=test.set)
test.pred <- predict(big.cart,newdata=test.set)
linear.oos <- lm(test.pred~test.set$Response)
int.oos <- linear.oos$coefficients[1]
slo.oos <- linear.oos$coefficients[2]
dev.new()
plot(x=test.set$Response,y=test.pred,main="CART Out-of-Sample
Predictions vs. Observed Responses",
     xlab="Observed",ylab="Predicted")
abline(a=int.oos,b=slo.oos)
#
dev.new()
plot(x=test.set$Response,y=test.pred,main="CART and Neural
Network Predictions vs. Observed Responses",
     xlab="Observed",ylab="Predicted",pch=1)
points(x=test.set$Response,y=net.pred.test,pch=2)
legend(x=33,y=52.5,legend=c("CART","Neural Net"),pch=c(1,2))
```

1.2.1.15.3 Code 18.3

```
setwd("Your path here")
#
df1 <- read.csv("20240806 Chapter 18 Neural Network Data.csv")
#
# Variables:
# X0-X9, Unit, Group, Response, Train.Data
#
```

```
#
#
#
library(randomForest)
library(nnet)
library(rpart)
attach(df1)
train.set <- subset(df1,subset=Train.Data==1)
test.set <- subset(df1,subset=Train.Data==0)
detach(df1)
attach(train.set)
big.forest <- randomForest(Response~X0+X1+X2+X3+X4+X5+X6
+X7+X8+X9)
big.net <- nnet(Response~X0+X1+X2+X3+X4+X5+X6+X7+X8+X9,size=4,dat
a=train.set,skip=TRUE,linout=TRUE)
net.pred <- predict(big.net,newdata=train.set)
big.cart <- rpart(Response~X0+X1+X2+X3+X4+X5+X6+X7+X8+X9,data=
train.set)
#
rsq.rpart(big.cart)
#
dev.new()
plot(big.cart,xlim=c(-0.1,12),ylim=c(-0.1,1.15),main="CART
Regression Tree")
text(big.cart,use.n=TRUE)
cart.pred <- predict( big.cart,newdata=train.set)
dev.new()
linear.model <- lm(cart.pred~train.set$Response)
int <- linear.model$coefficients[1]
slo <- linear.model$coefficients[2]
plot(x=train.set$Response,y=cart.pred,main="CART Predictions vs.
Observed Response",
     xlab="Observed",ylab="Predicted")
abline(a=int,b=slo)
detach(train.set)
attach(test.set)
net.pred.test <- predict(big.net,newdata=test.set)
test.pred <- predict(big.cart,newdata=test.set)
forest.pred <- predict(big.forest,newdata=test.set)
linear.oos <- lm(test.pred~test.set$Response)
int.oos <- linear.oos$coefficients[1]
slo.oos <- linear.oos$coefficients[2]
dev.new()
plot(x=test.set$Response,y=test.pred,main="CART Out-of-Sample
Predictions vs. Observed Responses",
```

```
     xlab="Observed",ylab="Predicted")
abline(a=int.oos,b=slo.oos)
#
dev.new()
plot(x=test.set$Response,y=test.pred,main="CART, Neural Net,
Random Forest Predictions vs. Observed Responses",
     sub="Test Data",xlab="Observed",ylab="Predicted",pch=1)
points(x=test.set$Response,y=net.pred.test,pch=2)
points(x=test.set$Response,y=forest.pred,pch=3)
legend(x=33,y=52.5,legend=c("CART","Neural Net","Random
Forest"),pch=c(1,2,3))
```

1.2.1.16 Chapter 19

1.2.1.16.1 Code 19.1

```
setwd("Your path here")
df1 <- read.csv("20240808 Chapter 19 Time Series Data Sim.csv")
#
library(forecast) #Contains the function Arima()
#
#Variables:
#
# white.noise
# ar2
# ma2
#

attach(df1)
n.size <- nrow(df1)
time.index <- seq(from=1,to=n.size,by=1)
acf(white.noise,main="Autocorrelation Function - White Noise")
dev.new()
pacf(white.noise,main="Partial Autocorrelation Function -
White Noise")
dev.new()
acf(ar2,main="Autocorrelation Function - AR(2)")
dev.new()
pacf(ar2,main="Partial Autocorrelation Function - AR(2)")
dev.new()
acf(ma2,main="Autocorrelation Function - MA(2)")
dev.new()
pacf(ma2,main="Partial Autocorrelation Function - MA(2)")
```

```
wn.model <- arima(white.noise,order=c(0,0,0))
wn.fitted <- fitted(wn.model)
wn.diff <- white.noise - wn.fitted
#
ar2.model <- Arima(y=ar2,order=c(2,0,0))
ar2.fitted <- fitted(ar2.model)
ar2.diff <- ar2 - ar2.fitted
#
ma2.model <- Arima(y=ma2,order=c(0,0,2))#this is the model
indiocated by the Box-Jenkins approach
ma2.fitted <- fitted(ma2.model)
ma2.diff <- ma2 - ma2.fitted
#
plot(x=time.index,y=white.noise,main="Time Series Data -
White Noise",
     xlab="Day",col="black",pch=1)
points(x=time.index,y=wn.fitted,pch=2)
#
dev.new()
plot(x=time.index,y=ar2,main="Time Series Data - AR(2)",
     xlab="Day",col="black",pch=1)
points(x=time.index,y=ar2.fitted,pch=2)
#
dev.new()
plot(x=time.index,y=ma2,main="Time Series Data - MA(2)",
     xlab="Day",col="black",pch=1)
points(x=time.index,y=ma2.fitted,pch=2)
#
dev.new()
hist.wn <- hist(wn.diff,plot=FALSE)
hist.wn$density <- 100*hist.wn$counts / sum(hist.wn$counts)
plot(hist.wn,main="Histogram of Observed - Fitted: White Noise",
     xaxt="n",xlab="Diff = White Noise - Fitted",ylab="Percent",
     xlim=c(-1.5,1.5),ylim=c(0,50))
axis(side=1,at=seq(from=-1.5,to=1.5,by=0.1))
#
dev.new()
hist.ar2 <- hist(ar2.diff,plot=FALSE)
hist.ar2$density <- 100*hist.ar2$counts / sum(hist.ar2$counts)
plot(hist.wn,main="Histogram of Observed - Fitted: AR(2)",
     xaxt="n",xlab="Diff = ar2 series - Fitted",ylab="Percent",
     xlim=c(-1.5,1.5),ylim=c(0,50))
axis(side=1,at=seq(from=-1.5,to=1.5,by=0.1))
#
dev.new()
```

```
hist.ma2 <- hist(ma2.diff,plot=FALSE)
hist.ma2$density <- 100*hist.ma2$counts / sum(hist.ma2$counts)
plot(hist.wn,main="Histogram of Observed - Fitted: MA(2)",
     xaxt="n",xlab="Diff = ma2 series - Fitted",ylab="Percent",
     xlim=c(-1.5,1.5),ylim=c(0,50))
axis(side=1,at=seq(from=-1.5,to=1.5,by=0.1))
#
```

1.2.1.16.2 Code 19.2

```
setwd("Your path here")
df1 <- read.csv("20240808 Chapter 19 Transition Probability
Matrix.csv")
#
#
# VARIABLES:
# S1
# S2
# S3
# S4
#
attach(df1)
Amat <- as.matrix(df1)
Smat <- eigen(Amat)
Smat$values
Smat$vectors
Lamda <- diag(Smat$values)
#
# This will compute the stationary distribution (if it exists)
# of a finite, irreducible Markov Chain
# if the input matrix is a one-step transition matrix
# for that chain
#
n <- 600 #number of transitions to get to steady state
Lamdan <- diag(Smat$values**n)
Eig <- as.matrix(Smat$vectors)
EigInv <- solve(Eig)
Arecover <- Eig%*%Lamda%*%EigInv
An <- Eig%*%Lamdan%*%EigInv #The rows of An are identical, and
are the stationary
#                              distribution for the chain
Lamda
Eig
Amat
```

```
Arecover
n
An
#
# rows of An will be the stationary distribution if n is
large enough
#
```

1.2.1.16.3 Code 19.3

```
setwd("Your path here")
#
set.seed(13)
n.size <- 120
white.noise <- rnorm(n=n.size,mean=0.0,sd=0.5)
ar2 <- arima.sim(n=n.size,model=list(ar=c(0.81,-0.5)),sd=0.5)
ma2 <- arima.sim(n=n.size,model=list(ma=c(0.9,0.1)),sd=0.5)
ts.plot(ar2)
dev.new()
ts.plot(ma2)
ts.data <- data.frame(cbind(white.noise,ar2,ma2))
write.csv(ts.data,"20240808 Chapter 19 Time Series Data Sim.csv")
```

Index

S. A. Pardo, *Beyond the T-Test*, https://doi.org/10.1007/978-3-031-84479-9

GPSR Compliance
The European Union's (EU) General Product Safety Regulation (GPSR) is a set of rules that requires consumer products to be safe and our obligations to ensure this.

If you have any concerns about our products, you can contact us on

ProductSafety@springernature.com

In case Publisher is established outside the EU, the EU authorized representative is:

Springer Nature Customer Service Center GmbH
Europaplatz 3
69115 Heidelberg, Germany

www.ingramcontent.com/pod-product-compliance
Lightning Source LLC
Chambersburg PA
CBHW071610210725
29917CB00003B/64
9783031844782